VIROLOGY MONOGRAPHS

FOUNDED BY

S. GARD, C. HALLAUER, K. F. MEYER

EDITED BY

D. W. KINGSBURY, H. ZUR HAUSEN

18

SPRINGER-VERLAG

WIEN NEW YORK

BIOLOGY
OF POXVIRUSES

BY

S. DALES AND B. G. T. POGO

SPRINGER-VERLAG

WIEN   NEW YORK

Professor SAMUEL DALES

Cytobiology Group, Department of Microbiology and Immunology,
Faculty of Medicine, Health Sciences Centre,
University of Western Ontario, London, Ontario, Canada

Professor BEATRIZ G. T. POGO

Departments of Experimental Cell Biology and Microbiology,
Mount Sinai School of Medicine, The Mount Sinai Hospital,
New York, N.Y., U.S.A.

With 27 Figures

Library of Congress Cataloging in Publication Data. Dales, S. (Samuel), 1921—. Biology of poxviruses. (Virology
monographs; 18.) Bibliography: p. 1. Poxviruses. I. Pogo, B. G. T. (Beatriz G. T.), 1932—. II. Title. III. Series.
[DNLM: 1. Poxviridae. W 1 VI 83 v. 18 / QW 165. 5. P 6 D 139 b.] QR 360. V 52 no. 18 [QR 412] 576'. 64 s. 81—9370

ISSN 0083-6591

ISBN-13:978-3-7091-8627-5   e-ISBN-13:978-3-7091-8625-1
DOI: 10.1007/978-3-7091-8625-1

Foreword

This volume, *Biology of Poxviruses*, marks our debut as editors of this well known series. We plan to continue the tradition of providing a forum for extensive, critical reviews of individual virus groups, as exemplified by the present volume.

But the pace of discovery is accelerating so rapidly that we feel the need to offer an additional format: volumes that contain collections of shorter, topical reviews on a group of related subjects. Such collections might cut across conventional boundaries between virus groups, dealing, as an example, with a particular aspect of virus-cell interaction.

Admittedly, this new format stretches the term "monograph" beyond the accepted definition, but we believe that we should pay that price to maintain the usefulness of the series as a medium of scientific communication.

Whenever possible, we will enlist the aid of deputy editors to bring such collections to fruition. As in the past, the editors and the publisher will welcome suggestions for topics and contributions.

D. W. Kingsbury   H. zur Hausen

Contents

I. Introduction

A. Scope of the Presentation

Smallpox, once a global infectious disease of man, has now been eradicated, due to intensive efforts sponsored and supervised by the World Health Organization. However, wide interest in the poxviruses persists, as attested by numerous articles published over the last 10 years. Several reasons may be offered to account for the continuing preoccupation with these agents. From the point of view of human disease, there is a close relationship between the agents of smallpox i.e., *variola major* and *minor*, and other mammalian viruses which are enzootic and which produce infection in man simulating variola (ESPOSITO *et al.*, 1977a, BAXBY, 1977a; ANDREWES and PEREIRA, 1974), raising the possibility that smallpox virus might re-emerge from one of these agents by mutation.

Regarding another disease of primary concern to dermatologists, is the close pathological similarity between epidermal nodules produced by the agent of *Molluscum contagiosum* and benign tumors formed in monkeys by infection with Yaba virus or in rabbits by fibroma viruses. This relationship continues to draw the attention of virologists investigating the general problem of cellular transformation and malignancy.

Furthermore, the high order of complexity associated with the structure, development, and function of the poxviruses has broadened interest in these agents as models of gene expression and control in eukaryotic cells. For such studies experimentalists now have at their disposal updated or new microtechnology for manipulating cell-virus systems, and they frequently employ a multidisciplinary approach combining genetics, biochemistry, immunology and electron microscopy.

We shall attempt to cover for the reader in considerable detail developments in poxvirus biology. To do justice to the earlier literature we have directed the reader, whenever possible, to suitable review articles. We are aware that, in developing this overview, personal bias may have resulted in an emphasis on some items of information in preference to other items, and these instances may appear to the *cognoscenti* as capricious or arbitrary selections of the cited work. But our consistent intention was to provide, as far as possible, a balanced and succinct survey of the biology of these agents.

B. Historical Development

Although smallpox has been known for centuries as one of humanity's most prevalent and deadly infectious diseases, the viral nature of the agent responsible was not understood until the turn of the present century. Yet, paradoxically, smallpox was among the first diseases shown to be amenable to prophylaxis, the means of prevention having apparently been discovered independently on different continents (LANGER, 1976). In Africa and Europe, practitioners of folk medicine inoculated fresh material taken from the pustule of a patient into skin abrasions to produce immunity. In ancient China, the practice of variolation prevailed, a procedure in which dried material prepared from pustules of patients recovering from a relatively mild disease was administered by inhalation to individuals, usually groups of children from the priviledged classes, kept in isolation under the supervision of physicians. When, at the instigation of the more enlightened eighteenth century leaders, inoculation in Europe and North America came into use as a formal routine in medical practice, in the best circumstances the protection afforded was incomplete, with refractory mortality ranging from one to several percent among inoculated individuals. Moreover, since care was not taken in the urban setting to isolate recently inoculated subjects from the susceptible population at large, the intended prophylaxes actually initiated local epidemics of smallpox among the contacts. Such experiences discouraged the acceptance of routine inoculation against smallpox as a mandatory public health measure. The advent of modern concepts of smallpox prophylaxis is properly dated from the practice of vacciniation (vacca being the latin word for cow), which commenced with the pioneering discovery of EDWARD JENNER, reported by him in 1798 (LANGER, 1976). JENNER realized that the variolae (pustules) which develop on the skin of cows contain a material (now known to be cowpox virus) capable of inducing a localized, mild disease in the human. In part, JENNER's conclusion was derived from observing that country milkmaids, who frequently were in contact with variolae of cows, were spared from smallpox. Today, we understand that the serological relatedness of the cowpox and variola agents explains the immunological protection afforded by the former. Now that variola has been eliminated as a human disease and the practice of mandatory vaccination is being discontinued, can one be certain that a smallpoxlike human agent might not again emerge? We shall deal with this matter in section IX (Genetics).

Concerning the physical nature of the poxviruses, BUIST has been credited with the first description, in 1887, of the elementary body (EB) or particle of vaccinia virus (see BLAND and ROBINOW, 1939). Nevertheless, it is VON PASCHEN, who described the EBs precisely in his initial (VON PASCHEN, 1906) and subsequent publications, to whom the first microscopic characterization of a poxvirus (Fig. 1) is frequently attributed. A correlation between the EB and the infectious entity was not, however, established until LEDINGHAM (1931) demonstrated that antisera produced against vaccinia virus or fowlpox virus could simultaneously agglutinate and neutralize the infectivity of EBs. Other evidence, provided by WOODRUFF and GOODPASTURE in 1929 (see SMADEL and HOAGLAND, 1942), showed infectiousness of fowlpox particles termed Borrell bodies. While these data were less incisive, they substantiated LEDINGHAM's discovery. The infectious Borrell

EBs become embedded during the occluded phase within proteinaceous cytoplasmic masses referred to as Bollinger bodies (BOLLINGER, 1873), presumably equivalent to more recently characterized acidophilic A-type inclusions (KATO et al., 1962a).

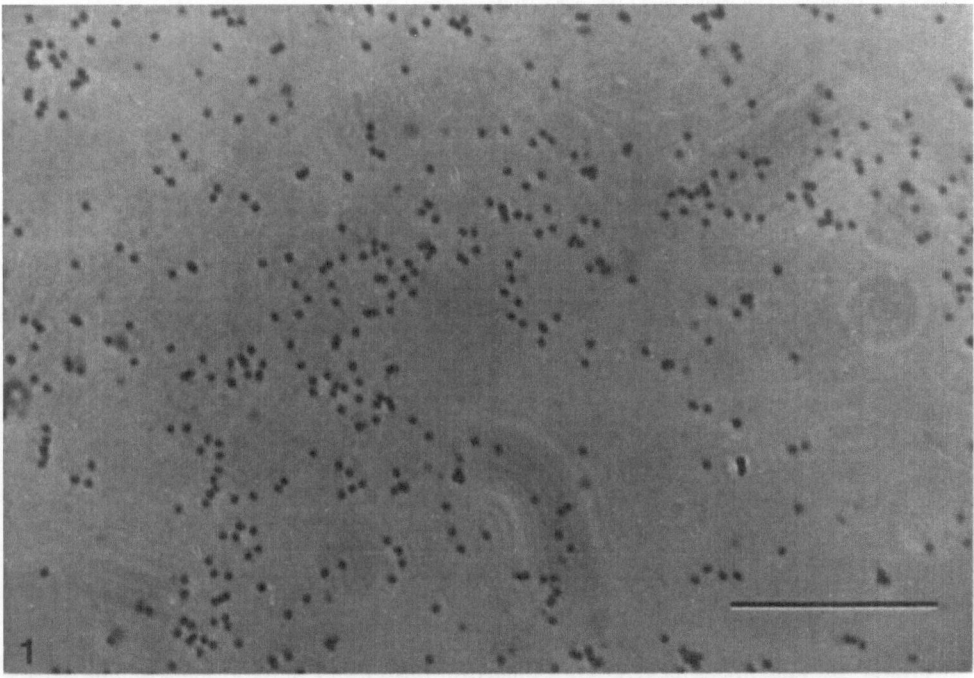

Fig. 1. Imprint of a rabbit cornea, infected with vaccinia virus, prepared by VON PASCHEN using the staining procedure developed by him. The slide was donated by VON PASCHEN to C. F. ROBINOW, Dept. of Microbiology and Immunology, University of Western Ontario, who kindly photographed a selected area in 1979, using bright field optics. Note the large number of uniform elementary bodies which occur in the vicinity of the infected tissue. × 3200. Bar is 1 μm in length

Intracellular structures associated with poxvirus infections were also identified at the turn of the century in both living and preserved tissue preparations. VOLPINO (1907) was able to observe the Brownian motion of EBs in living corneal tissue. Even earlier, GUARNIERI (1892) had suggested a relationship between the cytoplasmic inclusions which bear his name (and which are currently termed virus "factories" of CAIRNS (1960), where DNA replication and virus assembly occur) and a stage in the development of EBs. As the resolving power and image definition of the light microscope were improved, new techniques of specimen preparation were conceived that provided more accurate data about the poxviruses that remain valid to this day. Notable among these achievements was the use of annular oblique illumination by HIMMELWEIT (1938) to describe formation of so-called Marchal bodies in living cells of the chorioallantoic membrane infected with ectromelia virus. These Marchal bodies, which are equivalent to Bollinger

I. Introduction

Table 1. *Classification and salient*

Genus	Orthopoxvirus	Avipoxvirus	Capripoxvirus	Leporipoxvirus
Prototype of group	Vaccinia group	Fowl pox group	Sheep pox group	Myxoma group
Other members (host species)	buffalopox virus (buffalos) camelpox virus (camels) cowpox virus (bovines, man) ectromelia virus (mice) monkeypox virus (monkeys, man) rabbitpox virus (rabbits) variola virus (man)	canarypox virus juncopox virus pigeonpox virus quailpox virus sparrowpox virus starlingpox virus turkeypox virus	goatpox virus lumpy skin disease (Neethling) virus	hare fibroma virus rabbit (Shope) fibroma virus squirrel fibroma virus
Salient characteristics	Viruses of mammals. Infectivity of virions is insensitive to ether. Variable serological cross-reactivity and genome homology occurs. A hemagglutinin induced during infection appears at cell membranes but is not a component of the virion itself. Some members induce type A inclusions.	Viruses of wild and domestic birds. Infectivity is insensitive to ether. Serological cross-reactivity occurs between members of the group. Type A inclusions are induced but hemagglutinin is absent.	Viruses of ungulates. Infectivity is sensitive to treatment with ether. Serological cross-reactivity occurs between members of the group. Hemagglutinin is absent.	Serological cross-reactivity occurs between members of the group. Hemagglutinin is absent. Induce benign tumors.

[a] Derivation of names

pox: from plural of pock (Old English *poc*, *pocc-*) 'pustule, ulcer'
ortho: from Greek *orthos* 'straight, correct'
avi: from Latin *avis* 'bird'

capri: from Latin *caper*, *capri* 'goat'
lepori: from Latin *lepus*, *leporis* 'hare'
para: from Greek *para* 'by the side of'
entomo: from Greek *entomon* 'insect'

bodies or A-type inclusions of KATO *et al.* (1962a), described above, could be distinguished by HIMMELWEIT from EBs. Application of the dark-field technique coupled with U. V. illumination and improved lens systems permitted BARNARD (1931) to recognize clearly and measure individual EBs of ectromelia virus obtained from extracts of infected mouse tissues. He was also able to identify, by their similar appearance and size, the EBs within cytoplasmic MARCHAL inclusions and show, in collaboration with ELFORD, that the infectious component that filtered through collodion membranes of predetermined porosity was in fact the EB (appendix in BARNARD, 1931). Unfortunately, BARNARD was misled by images of paired EBs of ectromelia virus into the belief that he had observed a microorganism undergoing binary fission and his estimates of the sizes of the poxviruses were much too low.

Despite the early discoveries cited above and efforts by numerous other workers, precise data concerning the physical-chemical characteristics of the poxvirus EB and the nature of its replication did not begin to emerge clearly until the period commencing in the later 1930's and early 1940's. Among the most notable contributors belonging to the modern era of poxvirus research were CRAIGIE and his colleagues at the University of Toronto (CRAIGIE and WISHART, 1936a, b; WISHART and CRAIGIE, 1936), BLAND, ROBINOW and their associates in England (1939) and the Rockefeller Institute team in New York, comprised of SMADEL, HOAGLAND, RIVERS, SHEDLOVSKY, PARKER and others (SMADEL and HOAGLAND, 1942). The inevitable interrelationship between development of technology and rapid scientific progress was once again in clear evidence during this fertile period. Spectrophotometry permitted more sensitive microchemical

features of the Family Poxviridae [a]

Biologically related to *Lepori poxvirus?*	Parapoxvirus	Entomopoxvirus (insect species)		
Biologically like the Myxoma group?	Orf group	*Melolontha melolontha* subgroup A of Coleoptera	*Amsacta moori* subgroup B of Lepidoptera	*Chironomus luridus* subgroup C of Diptera
Molluscum contagiosum virus (human) swinepox virus (pig) Tanapox virus (human) Yaba monkey tumor poxvirus (monkey)	bovine pustular stomatitis virus chamois contagious ecthyma virus milker's node virus	*Anomala cuprea* *Aphodius tasmaniae* *Demodema bonariensis* *Dermolepida albohirtum* *Figulus sublaevis* *Geotrupes sylvaticus* *Othonius batesi* *Phyllopertha horticola*	*Acrobasis zelleri* *Choristoneura biennis* *Choristoneura conflicta* *Choristoneura diversana* *Chorizagrotix auxiliaris* *Operophtera brumata* *Oreopsyche angustella* and from the Orthoptera: *Melanoplus sanquinipes*	*Aedes aegypti* *Camptochironomus tentans* *Chironomus attenuatus* *Chironomus plumosus* *Goeldichironomus holoprasinus*
Serological cross-reactivity occurs between some members of the group. Cause benign tumors in primates and pigs.	Viruses of ungulates which occasionally infect man. The morphology of the virion is distinctively different from that of the other poxviruses of the mammals and birds (see section on "the Virion", below). Serological cross-reactivity occurs between members of the group. Hemagglutinin is absent.	Several different morphological types have been identified. The surface is generally covered with globular, rather than tubular elements of the warm blooded hosts. In some types only one, rather than two lateral bodies, is present. There is no serological cross-reactivity between viruses in this group and vertebrate poxviruses.		
		The virions are vivid and possess a single lateral body in association with a unilaterally concave core.	The virions are ovoid and posses a single lateral body in association with a cylindrical core.	The virions are brick-shaped like the vertebrate poxviruses, possess two lateral bodies and a bioconcave, plate-like core.

analyses and ultracentrifugation made it possible to concentrate and purify virus particles obtained from cow lymph or from rabbit epidermis in quantities of up to several hundred milligrams (LEDINGHAM, 1931; CRAIGIE, 1932). Methods were developed for sensitive immunological typing, for quantitation of infectivity using the REED and MUENCH (1938) end-point assay and for accurately determining the size, shape, number and purity of particles, taking advantage of the high resolving power of the electron microscope (VON BORRIES *et al.*, 1938; GREEN *et al.*, 1942).

Utilizing cytochemical methods adapted for FEULGEN staining of DNA, BLAND and ROBINOW (1939) documented in a synchronous infection that EBs of vaccinia become incorporated into the host-cell, then lose their identity, and then later induce the formation of large DNA-containing cytoplasmic inclusions within which developmental forms of progeny EBs appear. This study drew attention to the eclipse phase of the virus and the obligatory involvement of DNA-rich inclusions in the replicative cycle, alluded to previously in the 1914 publication of VON PROWAZEK (cited by BLAND and ROBINOW, 1939). More modern studies using either light or electron microscopy, reviewed by DOWNIE and DUMBELL (1956), JOKLIK (1966), and MOSS (1974), have corroborated the essential features of the studies by BLAND and ROBINOW.

By applying immunological procedures to disrupted EBs, CRAIGIE and colleagues distinguished a heat labile (L) from a heat stable (S) surface antigen of vaccinia virus, and both of these from a third, so-called nucleoprotein antigen. This work provided a clear demonstration of the antigenic complexity of the EB itself. All 3 components were also present as soluble antigens in preparations of dermal extracts employed as a vaccine (CRAIGIE and WISHART, 1936a, b). These

"LS agglutinogens" could elicit the formation of specific rabbit antibodies capable of agglutinating EBs or soluble antigens in the vaccine fluid and neutralizing infectivity (WISHART and CRAIGIE, 1936).

Having the capability of producing large quantities of EBs, and then being able to purify, concentrate and quantitate them as physical particles and infectious units enabled the Rockeller Institute team to generate a great deal of fundamental information about the chemical and physical characteristics of vaccinia virus (SMADEL and HOAGLAND, 1942) that provided a firm foundation for later studies.

C. Natural Distribution and Classification

Poxviruses have been identified in a large variety of wild and domestic mammals, in birds, and in four orders of insects (MATTHEWS, 1979; ANDREWES and PEREIRA, 1972; BERGOIN and DALES, 1971; GRANADOS, 1973). In the homoiotherms the predominating type of *Orthopoxvirus* is morphologically typified by vaccinia virus (Fig. 2). Parapoxviruses possess a more ovoid or elongated form (Table 1), illustrated in Fig. 3, as exemplified by contagious pustular dermatitis (Orf) virus of sheep and pseudo-cowpox (milker's node) virus. Among the more than 20 insect poxviruses identified to date there exists a greater variability in size and shape, as well as external and internal morphology (BERGOIN and DALES, 1971, GRANADOS, 1973a) (Table 1; Figs. 6, 13). However, in some insect agents, the presence of a biconcave core and two lateral bodies gives the virus an internal appearance that is virtually indistinguishable from that of vaccinia virus. To facilitate an objective identification of individual isolates among the numerous pro- and eukaryotic agents, the International Committee on Taxonomy of Viruses has adopted a classification scheme for each of the major virus groups including the poxviruses (MATTHEWS, 1979). In Table 1, which presents an abbreviated version of the most recently published classification (MATTHEWS, 1979), the poxviruses are grouped as a *Family* and further subdivided into *Genera* and *Species*. However, unlike species of plants and animals, which can be defined and grouped according to an objective Linnean classification, viruses by virtue of their obligatory parasitism can become adapted to replication in several and even many hosts. In some instances, exemplified by the wound tumor virus of clover, multiplication in both plant and insect hosts can take place (HSU *et al.*, 1977). Obviously, use of the Linnean classification scheme to identify viruses may be useful for some purposes, as in grouping antigenically related types (ESPOSITO *et al.*, 1977a—c; BAXBY, 1977a, b; MATTHEWS, 1979), or agents with nucleic acid homology (ARCHARD and MACKETT, 1979; MACKETT and ARCHARD, 1979; WITTEK *et al.*, 1980a) and for communicating information among virologists, but is not suitable for relating certain kinds of biological criteria of the type used to define *bona-fide* species of organisms.

II. The Virus Particles-Elementary Bodies

A. Isolation and Purification

Before the availability of tissue culture systems, vaccinia virus was routinely propagated in chicken embryos, notably on the chorioallantoic membrane (JOKLIK, 1962c), or it was obtained from calf lymph, the usual source of human vaccine material. Procedures were introduced by CRAIGIE and his colleagues (CRAIGIE, 1932) for laboratory production of EBs on a larger scale, to give yields up to 100 mg or more of virus particles. For this purpose, shaved areas of rabbit skin are inoculated by scarification, allowing the virus to proliferate and spread throughout the epidermis. Within a few days, the entire epidermal surface becomes a mass of dead cells filled with EBs, which can be collected readily by scraping the surface. The Rockefeller Institute group, using a combination of CRAIGIE's and LEDINGHAM's (1931) partial purification methods, developed a scheme for obtaining pure EBs (SMADEL and HOAGLAND, 1942). Briefly, rabbit epidermal pulp was suspended in distilled water, and the larger debris was removed by centrifugation at low g forces. Then, the virus was sedimented at higher speeds in an angle rotor of a centrifuge which had just been invented and developed by PICKELS. The homogeneity of EB preparations was established using electrophoretic analysis and sedimentation in an analytical ultracentrifuge. In both instruments, the virus particles formed a characteristic single boundary pattern, and examination in the electron microscope revealed the presence of isolated EBs free of contamination (SMADEL and HOAGLAND, 1942). In more recent times, we and others have also employed the rabbit skin in conjunction with modern purification schemes for producing large quantities of pure rabbitpox virus and can attest to the usefulness of this procedure (ZWARTOUW, 1964; P. GOLD and S. DALES, unpublished).

Following the advent of reproducible and easily manageable methods for routinely propagating cells *in vitro* as fresh explants from embryos and tissues or as continuous, defined cell lines, the poxviruses could be produced in a more fastidious and controlled manner. However, initially, because of practical limits on the number of tissue culture cells which could be routinely handled, only relatively small yields of EBs were obtained. During earlier times, when use of cell cultures in animal virology first came into vogue, stationary cultures in bottles were employed exclusively. Subsequently, techniques were introduced for propagating continuous lines such as the HeLa cell strain in agitated suspension (JOKLIK, 1962a), or as extensive monolayers in cylindrical, slowly rotating bottles. Both culture systems, especially suspension cultures, when scaled up to a volume of ten or more litres, can provide sufficient EBs for analytical work (MOSS and ROSENBLUM, 1974; JOKLIK, 1962d), in quantities approximating those obtainable from rabbit dermal pulp.

Cell-associated EBs, which constitute the bulk of the virus at the end of the growth cycle, can be dispersed or divested of their host-derived wrapping membranes by subjecting the lysates to multiple cycles of freezing and thawing (GREEN *et al.*, 1942; JOKLIK, 1962d; Moss and ROSENBLUM, 1973), by homogenization, or by disruption with ultrasonic oscillations (STERN and DALES, 1974). The capacity of fluorocarbons such as Genetron to remove preferentially the bulk of cell debris from lysates of infected cells, leaving behind the virus in the aqueous phase, has also been utilized for rapid purification of these agents (EPSTEIN, 1958a, b; PFAU and McCREA, 1963). When fluorocarbon extraction is followed by sedimentation through CsCl density gradients, highly pure preparation of virus can be obtained (PLANTEROSE et al., 1962). More commonly in use today is the procedure of HOAGLAND et al. (SMADEL and HOAGLAND, 1942), adopted by JOKLIK (1962c), and ZWARTOUW and his colleagues (1962), which includes the initial removal of larger cell particulates followed by differential centrifugation to concentrate the impure virus into pellets. The suspensions of isolated EBs are finally sedimented into bands by centrifugation through sucrose density gradients. In our experience, small fragments of adventitiously attached membranes of host origin are not completely removed from the EBs, even following several cycles of centrifugation through sucrose gradients. By contrast, sedimentation through potassium tartrate gradients, first employed with viruses by McCREA and colleagues (McCREA et al., 1961), does yield very pure, viable EBs, which are completely free from extraneous membrane contaminants (STERN and DALES, 1974). Although tartrate causes clumping of EBs, the removal of residual tartrate solution can be effected either by centrifugation or by dialysis through a semipermeable membrane and the EB aggregates can be dispersed by ultrasonic oscillations (MULLER, 1974; GEISTER and PETERS, 1969; MOSS et al., 1975).

B. Structure and Physical Properties

Although the uniformity and approximate size of some pox viruses was ascertained by the earliest investigators employing the light microscope, the precise size and shape of these agents first became evident from examinations in the electron microscope. Both VON BORRIES and the RUSKAS (1938) and later GREEN and colleagues (1942) described the oval or brick shaped profiles of dehydrated, unstained EBs of vaccinia and also recognized the presence of dense internal structures, which were a regular feature of the EBs. The literature of that period contains frequent remarks about the comparable sizes of EBs and the *Rickettsiae*, from which it was correctly surmised that poxviruses must be endowed with great biological and structural complexity. After coating by a layer of evaporated metal film, additional topological features became apparent on the EBs. PETERS and his colleagues coupled controlled protease and DNAse enzymatic hydrolysis with metal-shadowing in a series of classical studies which clearly revealed for the first time many of the subviral components of vaccinia virions. These components are now referred to as the envelope, lateral bodies, and the core or nucleoid containing the DNA (PETERS, 1956; STOEKENIUS and PETERS, 1955). Information derived from such whole mount preparations was corroborated and substantiated by means of thin sections of EBs embedded in plastic (PETERS, 1956; EPSTEIN, 1958b; DALES and SIMINOVITCH, 1961). The same general features characterizing vaccinia virus are fundamental to all agents in this Family. More recently devised methods for improved visualization of free or plastic-embedded virions, notably application of freeze etching and staining with salts of the heavy metals tungsten and uranium, has enabled the identification of many additional features in the fine structure of EBs during the past twenty years. These features include the surface ridges or surface tubular elements (STE) illustrated in Figs. 2 to 4 (BERGOIN and DALES, 1971; DALES, 1962; WESTWOOD *et al.*, 1964). In the case of parapoxviruses such as Orf virus, the ridges form a single, continuous helix

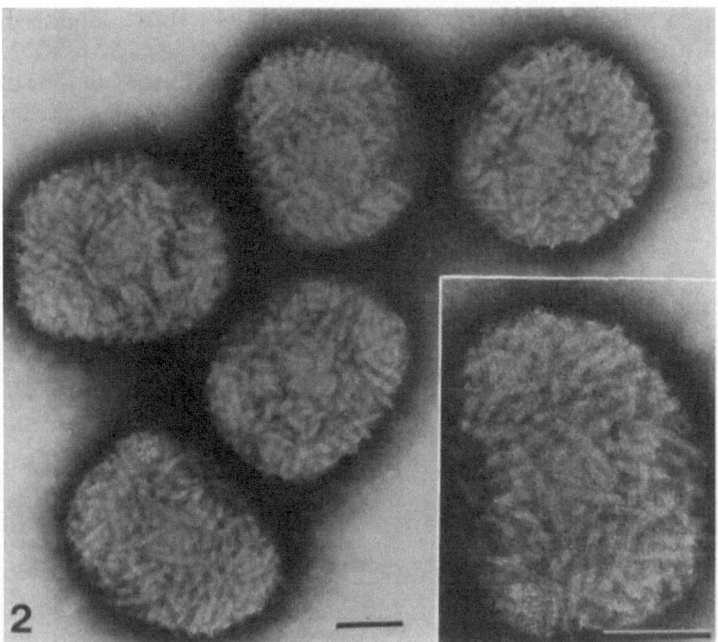

Fig. 2. Highly purified vaccinia virus examined as a whole mount after negative staining. The surface of each particle is covered by tubular structures (from GOLD and DALES, 1968). × 90,000; insert × 145,000. Bar is 1 μm in length

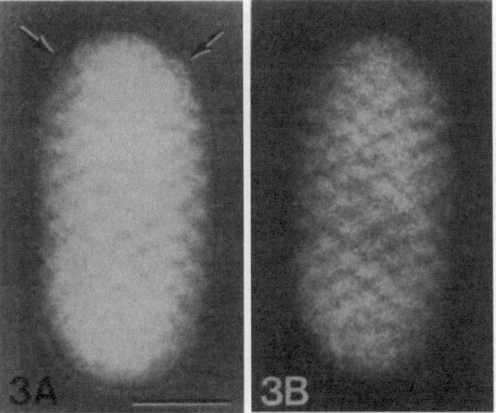

Fig. 3. Whole mount of Orf virus (pustular dermatitis) isolated from a human skin infection by L. HATCH and photographed by M. HALL (Dept. Microbiology and Immunology, University of Western Ontario). The continuous tubular surface structure is evident on both sides of the particle in *B*, giving the impression of a double helix (which actually does not exist). In *A*, the arrows indicate the limit of the envelope. × 135,000. Bar is 0.1 μm in length

at the virion surface. When the particles are viewed after staining with phos-
photungstic acid, images from both the top and bottom of the virion become
superimposed upon each other, giving rise to the crisscross pattern (PETERS *et al.*,
1964) illustrated in Fig. 3 B. The STE helix has been shown to be wound around
the virion in a left-handed sense (NAGINGTON *et al.*, 1964). Minute granules or
globular structures, illustrated in Fig. 4, can be seen at the edges of STE in replicas
or in freeze-cleaved preparations of vaccinia EBs (MEDZON and BAUER, 1970).
After vaccinia virus is heated to 50°, or in preparation of damaged virions, or
following treatment with non-ionic detergents such as Nonidet P40 or Triton
X-100, the convoluted appearance of normal envelopes evident by negative
staining changes. The envelope takes on a smooth or collapsed appearance, reveal-
ing, as in Fig. 5 A, the presence of an internal, rectangular core (EASTERBROOK,
1966; DALES, 1963). Some investigators designate damaged or denatured EBs
C forms and refer to normal virions as M forms (WESTWOOD *et al.*, 1964).

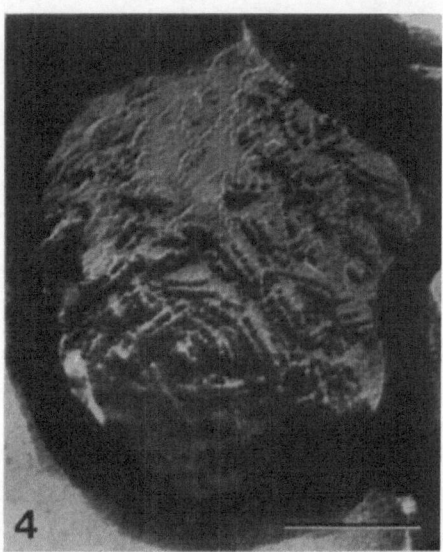

Fig. 4. Vaccinia virus examined as a replica of a whole mount, following freeze-etching.
Rows of small globules are evident along the edges of surface tubular structures.
(From MEDZON and BAUER, 1970, kindly provided by E. L. MEDZON, Dept. Microbiology and Immunology, University of Western Ontario.) × 179,000. Bar is 0.1 μm
in length

The internal structure of EBs is clearly evident in thin sections of embedded
virions, but details of fine structure, such as the pallisade layer of spicules surrounding the core, illustrated in Fig. 5 A, B, are revealed most clearly by negative
staining. Thinly sectioned EBs stained with uranyl salt solution reveal the
presence inside the core of DNA which sometimes appears as loosely tangled fine
threads or which may occur as tightly packed bundles of filaments (DALES, 1963).
In some preparations of vaccinia virus and other poxviruses, a folded coil or
cable 250 nanometers long and 40—50 nanometers in diameter has been observed

to occupy the space inside the core (Fig. 6). A dense 10 nanometer axis evident in longitudinal or cross sections at the centre of the cable was presumed to represent the genomic DNA surrounded by the less dense protein(s) of the cable (PETERS and MÜLLER, 1963).

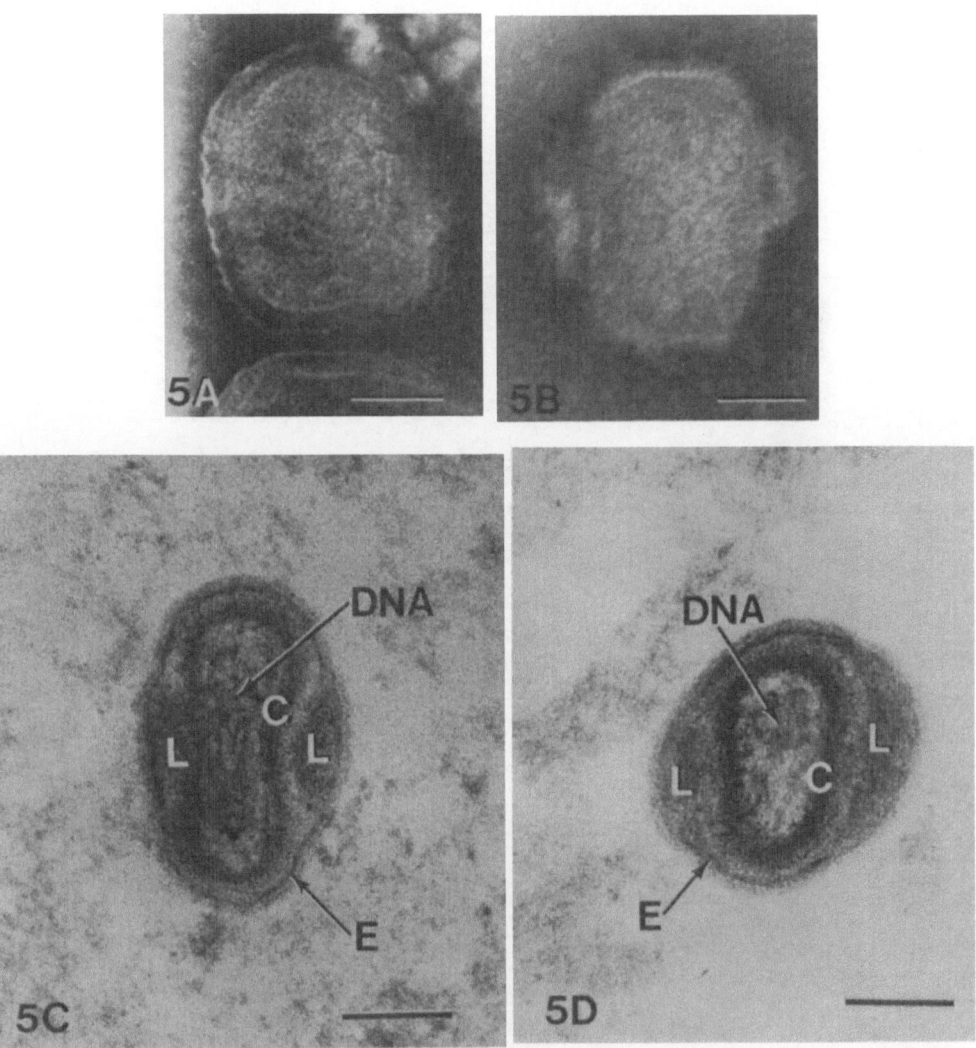

Fig. 5. *A* Negatively stained whole mount preparation of a vaccinia virion with a collapsed envelope. A short pallisade layer of spicules covers the central, rectangular core (from DALES, 1963). ×134,000

B An isolated core and lateral bodies evident in whole mount after negative staining. The envelope was removed by controlled degradation with NP 40 and 2-mercaptoethanol (From POGO and DALES, 1969a). ×120,000

C—D Thinly sectioned vaccinia virions illustrate the core (*C*) enclosing the dense, fibrillar DNA, two lateral bodies (*L*) and the envelope (*E*). C, section parallel and D, perpendicular to the long axis. ×140,000. The bars are 0.1 μm in length

In Table 2 are presented the dimensions of the salient features reported for representative poxviruses of the homoiotherms and insects. From these data it becomes clear that *parapox* viruses such as Orf virus, shown in Fig. 3, are relatively

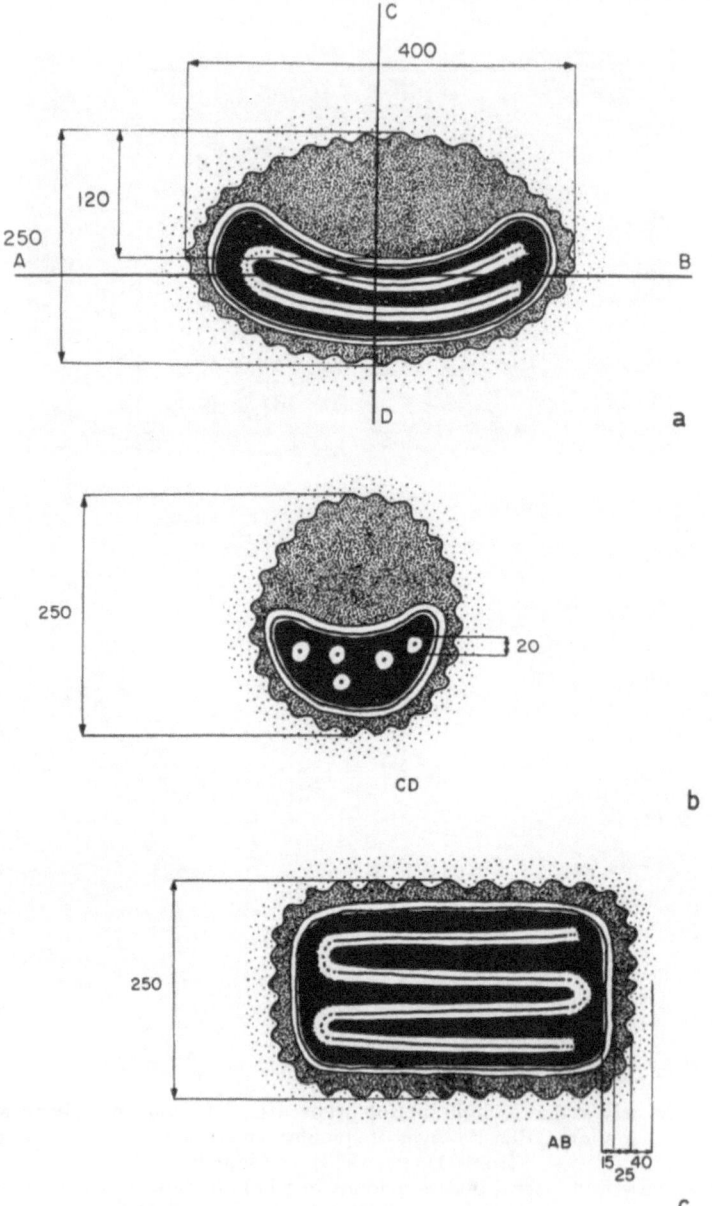

Fig. 6. Diagrammatic representation of an insect poxvirus, *Melolontha*, in the *a* longitudinal, *b* transverse and *c* horizontal planes of symmetry. Numbers are the average dimensions in nanometers. The cable structure is drawn at the center of the core (from BERGOIN *et al.*, 1971)

more elongated, with a 1.6 ratio of length to width. In comparison, *orthopox-viruses*, exemplified by vaccinia virus exhibit a length to width ratio of only 1.3 (BERGOIN and DALES, 1971; PETERS *et al.*, 1964; ROBERTS, 1968). The presence of one or two internal lateral bodies and their disposition in relation to the core also becomes a diagnostic feature of certain insect poxviruses, along with the folded cable structures within the core (Table 2 and Figs. 5B—D, 6).

Since the dimensions of EBs are ascertained from electron micrographs of particles prepared by a variety of procedures which inevitably cause, to a varying degree, damage or shrinkage due to dehydration, the size variability of 10—15% reported by different investigators is readily explained.

Table 2. *Sizes and characteristic features of the poxviruses*

Agent	Length mμ	Width mμ	Core shape and internal structure	Lateral bodies	References as listed in Bibliography
Vaccinia	235–280	165–225	biconcave, plate-like may contain folded cable structure	2 distinct	439, 487, 125 375, 122
Molluscum contagiosum	300	225	as above	as above	9
Fowlpox	330	285	as above	as above	9
Sheep pox	195	115	—	as above	9
Orf	250–295	160–190	as above	as above	9, 376
Chironomus midge (insects)	320	230	as above	as above	162, 452
Melolontha (insects)	400	250	unilaterally invaginated or kidney shaped, cable structure present	1 only lodged within the depression of the core	52, 162
Amsacta (insects)	350	250	prolate elipsoid	probably only 1 indistinctive surrounding the core	52, 162, 164

C. Composition and Physical Properties

Availability of pure EBs in adequate quantity enabled workers at the Rockefeller Institute to collect data on numerous physical and chemical properties of vaccinia virus. On exposure to various solute concentrations, the EBs were shown to respond to changes in osmotic pressure in a manner expected of entities enclosed by a semipermeable membrane (SMADEL and HOAGLAND, 1942). Thus, the density of EBs was found to change according to the composition of the suspending medium, being 1.16 g/ml in dilute buffer and 1.25 g/ml in 50% sucrose. In the analytical ultracentrifuge, the EB had a sedimentation coefficient of 4910s (s = Svedberg units). Assuming a brick shape for the virion and using the density and other physical measures, SMADEL and colleagues estimated the average dimensions of the hydrated EB to be 252 nm in length and 236 nm in width (GREEN et al., 1942), in good agreement with direct measurements obtained by investigators 20 years later using electron microscopy and negatively stained virions (WESTWOOD et al., 1964) (see Table 2). Weight estimations on suitably dehydrated preparations of pure vaccinia virions gave a value of 5.5×10^{-15} g per EB (SMADEL and LAVIN, 1940; SMADEL and HOAGLAND, 1942; JOKLIK, 1966), and for a parapoxvirus such as Orf virus, 3.7×10^{-15} g per EB (references cited in JOKLIK, 1966).

Table 3. *Chemical composition of vaccinia virus expressed as percentage of the dry weight*

Substance	References as listed in bibliography			
	438, 176, 439	560	203	
Principal components				
Nitrogen	15.3	14.7		
Phosphorus	0.57	0.49		
Sulfur	—	0.76		
DNA	5.6	3.2	5.25	
Cholesterol	1.4	1.2		
Phospholipid	2.2	2.1		
Neutral fat	2.2	1.7		
Trace material				
Carbohydrate	2.8	0.2		
Copper	0.05	0.02	0.02	
Riboflavin	1.3×10^{-3}	0.5×10^{-3}		
Biotin	present	1.3×10^{-5}		
RNA	trace	0.1	0.1	

Disruption of virions in alkaline solution enabled SMADEL et al. (1940) and HOAGLAND et al. (1940) to obtain extracts of nucleoprotein in which the nucleic acid was shown by spectrophotometry and colorimetric determinations to be of the thymus (i.e. DNA) type. Other components were also determined and the values obtained, shown in Table 3, are compared with those from more recent analyses by ZWARTOUW and others. An appreciably lower estimate of the DNA content found by ZWARTOUW (1964), is ascribed by this investigator to inaccuracies in the earlier estimations due to differences in calculating the mass of the EB, problems in extraction of the DNA, inherent errors in the colorimetric assays and other analytical errors. ZWARTOUW concludes that based on the phosphorus content of the virion, the lower estimate for percent of DNA is probably the more accurate. The physical-chemical characteristics of poxvirus DNA will be considered in detail subsequently.

Most investigators agree that RNA, if it is present in the virion at all is usually detectable in only trace amounts (SMADEL and HOAGLAND, 1942, ZWARTUOW, 1964, JOKLIK, 1966), perhaps due to incorporation during assembly of adventitious cellular RNA or virus-related transcripts which have not become dissociated from the DNA before packaging of the genome within cores (ROENING and HOLOWCZAK, 1974).

The presence of other substances in trace amounts suggests that they also are impurities which become adventitiously absorbed (ZWARTOUW, 1964). In the case of Cu++, the bound element is probably concentrated on the virus surface during purification (SMADEL and HOAGLAND, 1942), since EDTA is effective in its removal. Although carbohydrate occurs in only trace amounts, the presence of glucosamine in virion polypeptide(s) and other sugars in the envelope glycolipids makes these carbohydrates *bona fide* components of the virion (HOLOWCZAK, 1970; GARON and MOSS, 1971; MOSS et al., 1971; ANDERSON and DALES, 1978).

Lipid constitutes about 5% or more of the wet weight of the vaccinia virion (SMADEL and HOAGLAND, 1942; ZWARTOUW, 1964), but is claimed to comprise as much as 34% of the dry weight of the fowlpox virion (see LYLES et al., 1976 for additional citations). Approximately equal parts of phospholipid and cholesterol occur in vaccinia virions and both are totally extractable from the lipoprotein envelope by suitable lipid solvents or detergents (STERN and DALES, 1974; DALES and MOSBACH, 1968). While SMADEL and colleagues and subsequently others reported that removal of cholesterol by means of ether does not affect infectivity, ZWARTOUW (1964) observed that ether extraction at 37° did, in fact, reduce very appreciably the infectiousness of EBs. The phospholipids of the virion, while somewhat different in composition from those of the host cell in which the virus is propagated, are acquired from host cells (DALES and MOSBACH, 1968; STERN and DALES, 1974, 1976a). Host-derived glycolipids also occur in the virion (ANDERSON and DALES, 1978). In the case of fowlpox, the presence of squalene and cholesterol esters appears to be a unique finding (WHITE et al., 1968; LYLES et al., 1976).

D. Estimating Infectiousness of EB

The development of an accurate and reproducible quantitative measure of infectivity by means of the 50% end-point dilution or LD_{50} procedure of REED and MUENCH (1938) has enabled assays of the number of lethal doses for whole animals or for tissue culture tubes. Alternatively, assays of pock-forming units (PFU) can be conducted on rabbit skin (SMADEL and HOAGLAND, 1942) or on chicken chorioallantoic membranes (JOKLIK, 1966). More precise and easily reproducible titrations are possible by means of plaque assay in cell culture with either liquid overlay or under agar or methyl cellulose overlay.

The enumeration of total virus particles is carried out by enumerating in the electron microscope known concentrations of latex spheres of uniform diameter mixed with virus suspensions. Samples may be prepared by the microdrop atomizer spray technique of WILLIAMS and FRAZER (1956) or by deposition and dialysis through formvar membranes (SHARP, 1965), sometimes followed by negative staining (DALES, 1962), or by depositing all virions from a sample under uniform conditions by centrifugation onto surfaces from which replicas are produced (SHARP and BEARD, 1952; SHARP and McGUIRE, 1970). The number of EBs in a preparation may also be estimated from the mass of protein or dry weight of the sample, when sufficient material is available for analysis (SMADEL and HOAGLAND, 1942). Whichever procedure is adopted, the ratio of infectivity to particle number may be derived by relating the concentration of biologically active, infectious virus to the number of physical units present in a given preparation. When clumping is minimized, there may be as few as 2.4 EBs per infectious unit, although values of 4:1 to 10:1 are much more common. If clumping or some form of denaturation occurs, the ratio of EBs to infectious units may increase to 50:1 or even 1000:1 (SMADEL and HOAGLAND, 1942; DALES, 1962).

E. Virion-Associated Antigens

Early studies with purified EBs by CRAIGIE, and SMADEL and HOAGLAND (1942) could identify three principal antigenic determinants, a so-called heat labile (L) and heat stable (S) external antigen and an internal nucleoprotein (NP) antigen common to variola virus, vaccinia virus and many other homoiothermic poxviruses. Subsequent investigators, employing extracts of alkaline digests of vaccinia virus, were able to recognize at least 8 precipitin lines by the OUCHTER-LONY double-diffusion procedure in agar (ZWARTOUW et al., 1965) and by immuno-electrophoresis. Digests of rabbit poxvirus prepared from rabbit epidermal pulp and vaccinia virus solubilized by heating in sodium dodecyl sulphate (SDS) solution yielded analogous data (P. GOLD and S. DALES, unpublished). However, the total infected rabbit skin pulp or lysate of infected cultured cells is found to contain at least 17 identifiable antigenic determinants, some of which are non-virion but virus-specified functions (ZWARTOUW et al., 1965; WESTWOOD et al., 1965). One of the prominent antigens present at the surface, which can elicit antibody that neutralizes infectiousness, resides in the surface tubular elements (STE), as demonstrated by employing isolated purified STE protein as the immunogen (STERN and DALES, 1976b; DALES et al., 1976).

F. Virion-Associated Polypeptides

The technical advances resulting from the invention of electrofocusing by ampholytes and electrophoresis through acrylamide gels, enable one to separate individual polypeptides according to their net electric charge or molecular weight. In the 1970's, these advances produced a wealth of new information about the chemical constitution of viruses that parallels in scope the advances made by electron microscopy in the 1950's and 1960's.

The earliest analysis of SDS-dissociated EB polypeptides by means of poly-acrylamide gel electrophoresis (PAGE) was made by HOLOWCZAK and JOKLIK (1965), who used isotopically labelled virions to demonstrate at least 17 clearly defined polypeptide bands in cylindrical disc gels. The virus core was found to contain one major and several minor components. The nonionic detergent, NP40, could extract from the surface envelope layer two abundant polypeptides. When longer SDS-cylindrical gels came into subsequent use, 30 polypeptide bands were revealed by staining and/or by application of autoradiography to longitudi-nally sliced gels (SAROV and JOKLIK, 1972a, b). Similar polypeptide complexity and heterogeneity was demonstrated with fowlpox virus (OBIJESKI et al., 1973) and the Yaba monkey tumor agent (FENGER and ROUHANDEH, 1976). Using the slab discontinuous SDS-PAGE procedure of LAEMMLI (1970), 55 or more individual polypeptide species of vaccinia virus became evident (Fig. 7A; DALES et al., 1976). Judging by comparisons of PAGE analyses of vaccinia, cowpox, and Shope

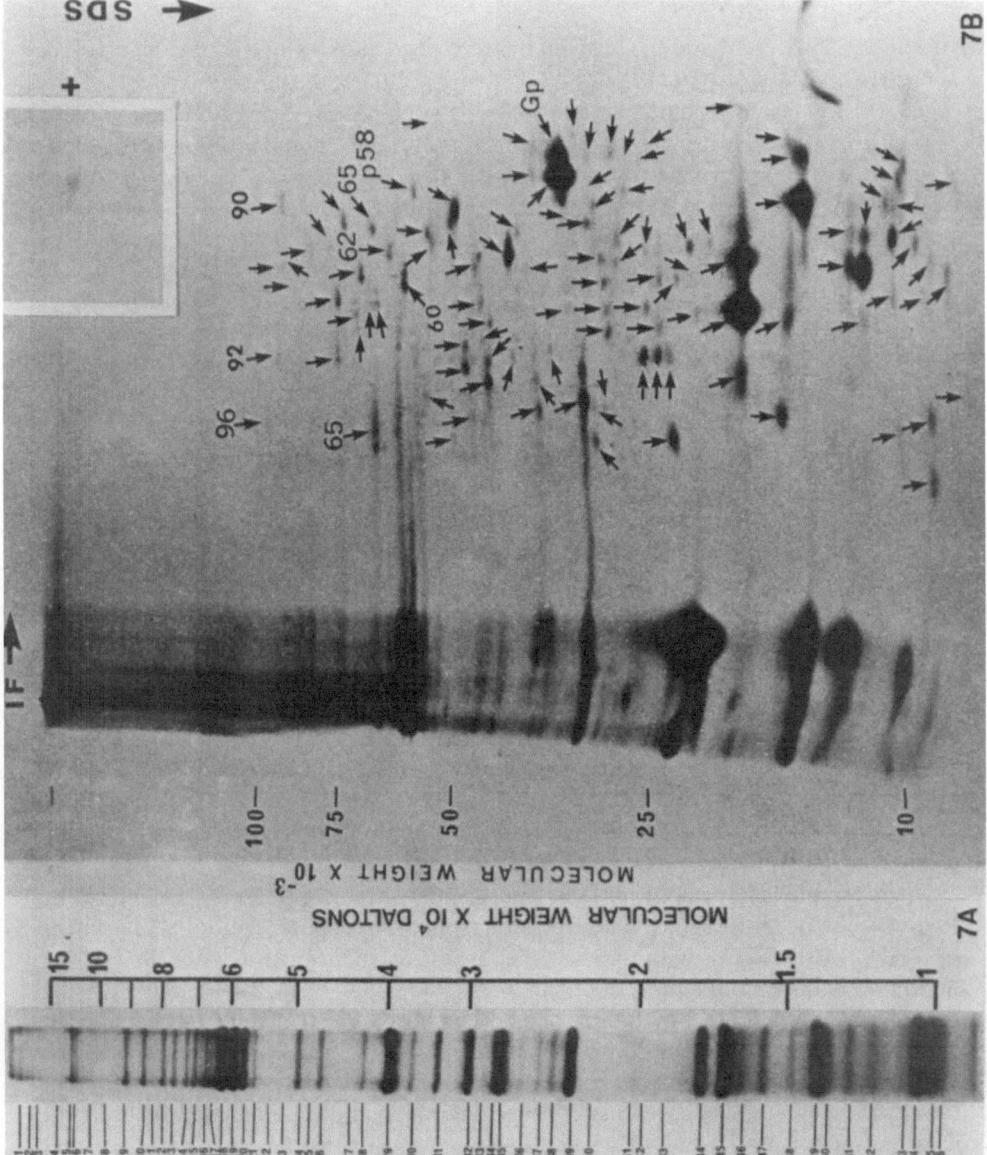

Fig. 7. *A* Autoradiogram of one-dimensional SDS-PAGE of [³⁵S]methionine-labeled pure vaccinia virus. All identifiable bands are numbered on the left and the molecular weight scale is shown on the right. About 20 µg of virus protein containing 20,000 cpm was applied to 11% acrylamide slab gels. The numbers on the left indicate the numerical order of the bands

B Autoradiogram of a two-dimensional separation of vaccinia virion polypeptides. The large panel is a preparation of whole virions, and in the inset, lower right, a sample of purified 58 K surface tubular elements (STE). Pure [³⁵S]methionine-labeled virus was dissociated and aliquots, each containing 15 µg of protein and 170,000 cpm, were introduced into isoelectrofocusing (IF) gels at the cathode. Electrophoresis was conducted for 6000 V · hr, following warming to 37° for 60 min to equilibrate the system. The IF gels were then placed onto slab gels and electrophoresis in the second dimension was carried out for 5 hours at 25 A. Arrows indicate individual polypeptides (from ESSANI and DALES, 1979)

fibroma virions, the band patterns indicate both homology and variability among the early and late proteins (IKUTA *et al.*, 1978a, b).

The 2-dimensional combination of isoelectric focusing and PAGE has already proven to be more informative regarding the number, modifications, and genetic variability of poxvirus-specific polypeptides (Fig. 7 B; ESSANI and DALES, 1979). The vaccinia virion was shown by this procedure to contain at least 110 polypeptide species, while similar analysis of extracts from infected cells revealed an even greater number of protein spots. For the sake of clarity, in discussing the data on molecular weight (MW) of polypeptides which were identified from various laboratories, we have abbreviated the numerical values by designating each 1000 daltons as 1 Kilo-; thus, MW 10,000 = 10 K and 100,000 = 100 K etc. The distribution of MW of vaccinia EB polypeptides observed by slab PAGE analyses ranges from 8 K to 200 K.

In order to identify the polypeptides that are accessible on or near the surface of EBs, SAROV and JOKLIK (1972a) tagged them *in situ* with the isotope I^{125} by means of the lactoperoxidase reaction or by labelling with fluorescein isothiocyanate. Five proteins identified in this manner were the same as those lost when the envelope was stripped away by EASTERBROOK's (1966) controlled procedure using NP 40 and chymotrypsin (Fig. 5 B). The virus core was shown to contain 17 polypeptides, among which the 60 K and 62 K species are the two principal ones. The virion contains 2 glycoproteins, 38—39 K and 41 K, each with carbohydrate chains containing glucosamine residues (Moss *et al.*, 1971). Neither glycoprotein is a constituent of the core or the envelope (HOLOWCZAK, 1970; GARON and Moss, 1971), contrary to initial conclusions about the putative surface location of these glycoproteins based on data from lactoperoxidase-I^{125} labelling. Thus, the envelopes of orthopoxviruses, unlike those of enveloped agents formed by 'budding', do not appear to contain any glycoproteins.

A major phosphoprotein of 11 K and several minor ones were discovered by SAROV and JOKLIK (1972a). The same phosphorylated polypeptide, representing over 11 % of the EB protein mass, was subsequently shown to be identical with a highly basic histone-like protein of the virus core (POGO *et al.*, 1975). The presence of 4 basic polypeptides, including the 11 K phosphoprotein and 3 others, 24 K, 34 K and 58 K in molecular weight, all amenable to extraction from the virion cores with sulfuric acid, has been reported (LANZER and HOLOWCZAK, 1975). The lower MW polyamines, spermine and spermidine, could be identified in both the core and elsewhere in the virion by labelling vaccinia virus with ^{3}(H)-ornithine during replication (LANZER and HOLOWCZAK, 1975).

Another very prominent polypeptide of the virion which occurs near the surface is the 58 K STE component which can be isolated in a pure state, as illustrated in Fig. 8 (STERN and DALES, 1976b).

G. Virion-Associated Enzymatic Activities

Although it had been known for many years that certain viruses possess enzymatic activities as in the case of the neuraminidase of influenza virus (HIRST, 1943), the first viral enzyme involved directly with virus genome expression to be identified was the DNA-dependent RNA polymerase within the core of a poxvirus. The almost simultaneous discovery of this enzyme by KATES and MCAUSLAN (1967c) and by MUNYON and colleagues (1967) provided a satisfactory explanation for the observed autoregulation by the inoculum of its own uncoating after penetration into the host cell cytoplasm. Previously, uncoating had been presumed to be completely under cellular control (JOKLIK, 1966). Undoubtedly, recognition of the poxvirus RNA polymerase established the precedent which culminated in the discovery of analogous polymerases in other animal viruses, among which the most notable were the RNA-dependent DNA polymerases of oncogenic and nononcogenic retroviruses (reviewed by TEMIN, 1971, and BALTIMORE, 1971).

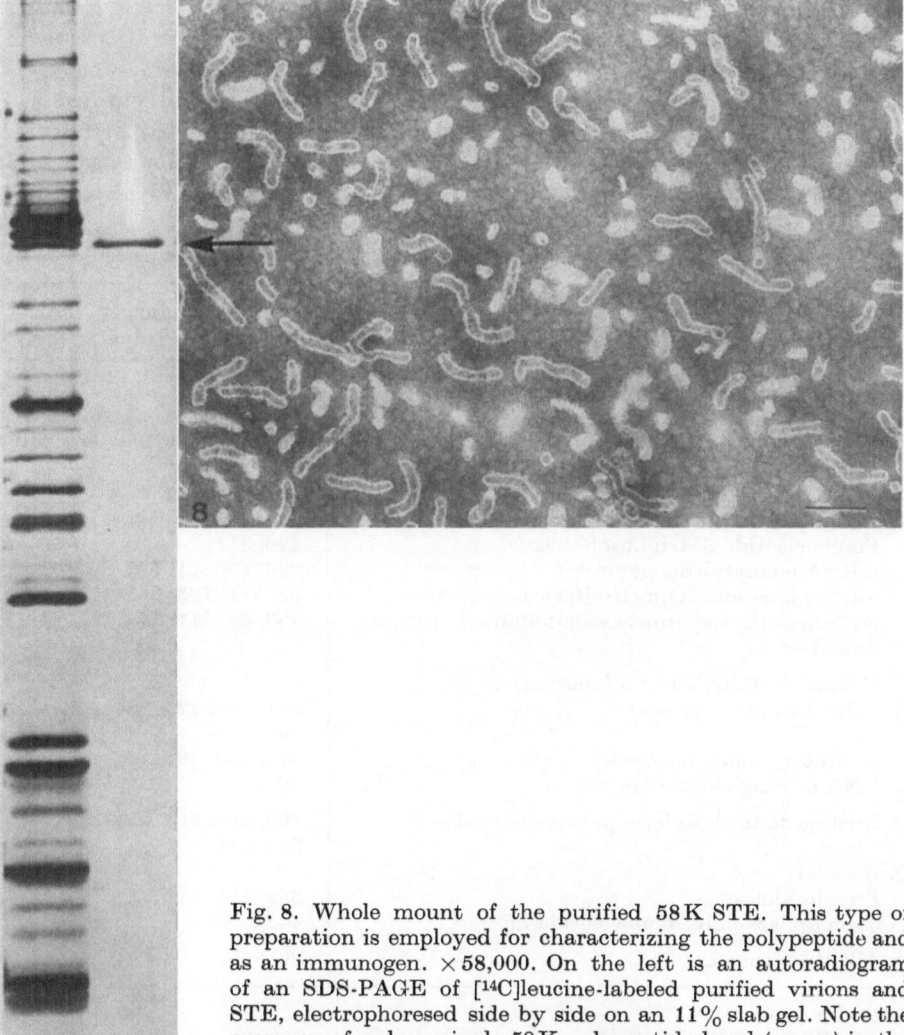

Fig. 8. Whole mount of the purified 58 K STE. This type of preparation is employed for characterizing the polypeptide and as an immunogen. ×58,000. On the left is an autoradiogram of an SDS-PAGE of [14C]leucine-labeled purified virions and STE, electrophoresed side by side on an 11% slab gel. Note the presence of only a single 58 K polypeptide band (arrow) in the channel loaded with STE. (From STERN and DALES, 1976)

During the 13 years which have passed since the DNA-dependent RNA polymerase was identified in vaccinia virus cores, similar transcriptases have been found in many of the homoiotherm and insect poxviruses (see Table 4). Furthermore, due to intensive efforts in several laboratories, agents belonging to this family have so far been shown to contain at least 12 or 13 other enzymatic activities (see Table 4 for references). We surmise that the presence of such a large spectrum of enzymes endows the poxvirus with a great deal of autonomy in control of the expression of its genome, particularly during the early phases of the infectious process, prior to uncoating (i.e., release of the DNA from the core). However, one should realize that the manifestation of a particular enzyme activity in a virus particle does not, *per se*, constitute sufficient proof that such an activity is virus-encoded, since spurious reactions may occur as a consequence of adventitious adsorption of non-virus enzymes from the cell or tissue extracts. Examples are the lipase and other enzymes in the purified vaccinia virus preparations studied by SMADEL and colleagues (SMADEL and HOAGLAND, 1942). Instances may also occur when an enzyme presumed to have been solubilized and purified as a distinctive polypeptide(s) may be contaminated by an unrelated activity, as appears to be the case with one of the vaccinia virus DNA exonucleases, to be described below (POGO and O'SHEA, 1977). Despite these reservations, it is

Table 4. *Enzymatic activities identified within the core of poxviruses*

Activity	References (as listed in Bibliography)
A. Related to transcription and modification of mRNA	
DNA-dependent RNA polymerase [a, c]	331a, 217, 316, 220, 423, 389, 197, 305, 349, 350, 509, 510, 214, 356
Endoribonuclease	308, 331, 257, 266, 305, 215
Polyadenylate polymerase(s) (terminal ribo-adenylate transferase) [c]	
Polynucleotide 5′-triphosphatase [c]	470, 311
mRNA guanylyltransferase [b, c]	276, 65, 311, 123
mRNA (guanine-7-)-methyltransferase [b, c]	65, 311, 123
mRNA methylase (nucleoside-2′-0-methyl transferase) [b, c]	483, 65, 311, 34
B. Related to DNA and its functions	
ss DNAse, exonuclease [a, c]	387, 386, 423, 389, 406, 407, 393
ss DNAse, endonuclease [a, c]	423, 389, 393
DNA nicking-closing enzyme [c]	39
C. Nucleoside triphosphate phosphohydrolase(s)	160, 317, 318, 423, 389, 351, 353, 354
D. Kinases	
Protein kinase [c]	352, 111, 246
5′-phosphate-polyribonucleotide kinase [c]	443
E. Alkaline protease	20

[a] late function
[b] early function
[c] partially or highly purified

satisfying to be able to report that almost all of the enzymes listed in Table 4 have been successfully separated from poxvirus cores and either partially or completely purified to levels whereby assignment of specific activities to individual virus-specific polypeptides or groups of polypeptides can be made.

Although some of the virion-associated enzymes provide functions required during the early stages of virus penetration, before synthesis of viral DNA, several are either synthesized or become detectable by their activities only after virus DNA replication (i. e. they are late functions) (Table 4). The activities of some of the late enzymes become manifest in a sequence of events temporally coordinated with post-translational cleavage of structural proteins and differentiation of the virion structure, implying that cleavages occur when virus cores are formed during the process of maturation (STERN et al., 1977). In the following description of the manner in which the core enzymes participate in synthetic events, we shall for the sake of clarity group their activities into classes of functions, as they are set out in Table 4.

i) Enzymes Related to DNA Transcription and Modifications of the Polyribonucleotide Messengers (mRNA)

Following the discovery of a DNA-dependent RNA polymerase activity within the core of vaccinia virus by KATES and McAUSLAN (1967a, c), and MUNYON et al. (1967), a series of new observations has been published concerning mechanisms for synthesizing, modifying, and extruding mRNA from the virus core. Attention has been given to core-associated transcription because the poxvirus offers material suitable for studying in great detail some mechanisms for expression of genetic information and provides a unique model for obtaining insights into transcriptional and post-transcriptional events of more general relevance to eukaryotic organisms.

The informational or nucleotide sequence homology contained in RNA transcripts synthesized by cores in vitro or within the host cell cytoplasm was first examined by KATES (1970). The in vitro and in vivo synthetic events show similarities with respect to regulation of the transcription process by which defined classes of early mRNAs are synthesized and modified post-transcriptionally. The virion RNA polymerase, like all the other core enzymes, can be activated by treatment of virus particles with non-ionic detergents, sometimes in conjunction with sulfhydryl reagents such as mercaptoethanol (EASTERBROOK, 1966), to cause the permeation or removal of the lipoprotein envelope (KATES, 1970; POGO et al., 1971; SCHWARTZ and DALES, 1971). Included in the basic reaction mixture employed in the original studies were the 4 ribonucleoside triphosphates, an ATP regenerating system, consisting of phosphoenolpyruvate and pyruvate kinase, a buffer to maintain the pH at ~9.0 and the divalent cation Mg^{++}, which could be partially replaced by Mn^{++}. In a more recent study, involving a vaccinia virus-induced RNA polymerase isolated from the cytoplasm, the pH optimum of the reaction was found to be 7.9 and an obligatory requirement for Mn^{++} was reported (NEVINS and JOKLIK, 1977a). The in vitro polymerizing reaction can be maintained at a maximum rate for longer than 2 hours, whereby large quantities of RNA may be synthesized (KATES, 1970; McRAE and SZILAGYI, 1975). Repeated attempts in the past to isolate and purify the RNA polymerase from virus cores have been unsuccessful. This failure may have been due to the complex nature of this polymerase, judging by the multimeric structure of the enzyme isolated from the cytoplasm of infected HeLa cells by NEVINS and JOKLIK (1977a). To recover and purify the cytoplasmic enzyme, use was made of column chromatography on DEAE-Sephadex and phosphocellulose, followed by precipitation with ammonium sulfate and sedimentation in isopycnic glycerol gradients. Analysis by poly-acrylamide gels revealed the presence of 7 polypeptides of MW (a) 135 K, (b) 130 K, (c) 77 K, (d) 34 K, (e) 19.5 K, (f) 16.5 K and (g) 13.5 K, respectively. These subunits occur in equimolar ratios. Polypeptides (a), (b) and (c) are of identical MW to the three polypeptides known to exist in the virion core and it is therefore presumed that the other polypeptides. (d) through (g), also exist in the core. The vaccinia RNA polymerase recently purified from virions (SPENCER *et al.*, 1980 and BAROUDY and MOSS, 1980 appended bibliography), does indeed possess the same properties as the enzyme isolated from infected cells. Judging by their MWs, none of the 7 polypeptides listed could be related to any of the polypeptide subunits of RNA polymerases I or II of the host HeLa cells, indicating that the vaccinia enzyme is entirely virus-specified. Applying information about the amount of protein contained in vaccinia virus RNA polymerase polypeptides (a), (b) and (c), pre-sumed to represent 2.5% of the virion protein mass, and assuming that the 120×10^6 dalton DNA of vaccinia virus constitutes about 5% of the mass of the virus, NEVINS and JOKLIK (1977a) calculated that each virus core should contain, on the average, 150 to 200 molecules of the complete enzyme. This estimate is in reasonably good accord with the previous estimate of 50 molecules per virion by KATES and BEESON (1970a), which was based on the rate of *in vitro* tran-scription from cores. Thus, the presence of a multitude of RNA polymerase molecules ensures that the process of transcription is rapid and efficient during the early phases of the infectious cycle. Calculations of abundance made on the other core activities indicate that some enzymes may be present in only a few copies, while others occur with a molecular frequency about as great as that of the transcriptase.

It has been generally reported that the RNA products of *in vitro* transcription from the cores have a size range of 8 to 14S. More recently it was established by PAOLETTI (1977a, b) that, immediately after short pulse-labelling, larger RNA products of 20 to 30S are evident within the core. Upon the appropriate chase intervals, in experiments which sometimes involved the application of analogues of ATP to interrupt both RNA synthesis and the extrusion of mRNA, the higher MW RNA was shown to be converted to lower MW mRNA coincident with extrusion from the core. PAOLETTI and LIPINSKAS (1978a) have obtained evidence suggesting that segmentation of the precursor into smaller mRNA classes is catalyzed by a core-associated endoribonuclease, an activity which in the solu-bilized state can specifically cleave the 20 to 30S RNA from vaccinia cores.

The presence of polyadenylic acid [poly(A)] tracts at the 3′ ends of many species of mRNA from the eukaryotes and their viruses is well established and the mRNAs of the poxviruses are not excepted. The early class of 8 to 14S mRNA transcribed from the virus core is modified at the 3′ end by covalent linkage to poly(A) chains comprising 50 to 200 residues. These are appended to the mRNA following a non-transcriptional synthesis (SHELDON and KATES, 1974). To synthesize the poly(A) tracts, the core contains a riboadenylate transferase, termed also poly(A) polymerase (KATES and BEESON, 1970b; Moss *et al.*, 1973; McKAY BROWN *et al.*, 1973; Moss and ROSENBLUM, 1974; SHELDON and KATES, 1974). An activity with properties very similar to those of the core-associated enzyme has been isolated from the cytoplasm of infected HeLa cells (NEVINS and JOKLIK, 1977b). The cytoplasmic enzyme has a 57K and a 37K polypeptide, optimum at pH 8.6, a preference for Mn^{++} as the divalent cation and no homology in terms of the MW of its subunits with corresponding nuclear or cytoplasmic poly(A) polymerases from HeLa cells. Calculations suggest that 100 or more molecules may be present per virion core (Moss *et al.*, 1975).

Apart from the addition of poly(A) stretches at the 3' end, virion-specified mRNA is also modified by "capping" at the 5' terminus. The "caps" formed in the *in vitro* synthesized product are $m^7G(5')pppA^m$ or $m^7G(5')pppG^m$, in which m^7G is 7-methylguanosine, whereas G^m and A^m are 2'-0-methylguanosine and α'-0-methyladenosine (WEI and Moss, 1974; MARTIN *et al.*, 1975; BOONE *et al.*, 1977; GERSHOWITZ and Moss, 1979). A multistep series of enzymatic reactions related to the "capping" phenomenon has been identified in association with the virus core by Moss, PAOLETTI and their colleagues (WEI and Moss, 1974; TUTAS and PAOLETTI, 1977; Moss *et al.*, 1976; BOONE *et al.*, 1977; ENSINGER *et al.*, 1975). The reactions have been postulated to follow the sequence described in Fig. 9 according to the scheme of TUTAS and PAOLETTI (1977), and Moss *et al.* (1976):

1. $\gamma'\beta'a'pppNpN \xrightarrow{(a)} \beta'a'ppNpN - +\gamma'Pi$
2. $\gamma\beta apppG + \beta'a'ppNpN \xrightarrow{(b)} G(5')a\beta'a'ppp(5')NpN - +\gamma\beta ppi$
3. $G(5')a\beta'a'ppp(5')NpN - + AdoMet \xrightarrow{(c)} m^7G(5')a\beta'a'ppp(5')NpN - + AdoHcy$
4. $m^7G(5')a\beta'a'ppp(5')NpN - + AdoMet \xrightarrow{(d)} m^7G(5')a\beta'a'ppp(5')N^npN - + AdoHcy$

AdoMet = S-adenosylmethionine
AdoHcy = S-adenosylhomocystene

Fig. 9. Sequence of reactions in vaccinia virus cores related to "capping" of mRNA

Reaction (a) in the scheme utilizes polynucleotide 5'-triphosphatase. The activity has been solubilized from cores and partially purified by means of poly(U)-agarose affinity chromatography (TUTAS and PAOLETTI, 1977). The enzyme is a heterodimer of MW 113K, consisting of 90K and 26K subunits. Activity depends on the presence of a divalent metal such as Mg^{++} and has a pH optimum of 8.4. The enzyme hydrolyzes the γPi on 5'-ATP or 5'-GTP-terminated heteropolymeric RNA, but also functions on 5'-ATP-terminated poly(A).

Reaction (b) involves GTP:RNA guanylyltransferase and reaction (C) the S-adenosylmethionine:mRNA(guanine-7-)-methyltransferase, abbreviated as 7-methyltransferase. Both activities have been solubilized (MARTIN and Moss, 1975; MARTIN *et al.*, 1975; BOONE *et al.*, 1977; Moss *et al.*, 1976), highly purified by MONROY *et al.* (1978a; b), and shown to reside in a complex of MW 127K which is a heterodimer of 95K and 31K polypeptides. Neither activity has been separated from the other or identified with specific polypeptide species. One subunit of the enzyme utilizes GTP as the guanylyl donor and functions specifically on 5'-triphosphate-terminated RNA chains. Both activities are apparently synthesized as early virus functions in the cytoplasm of infected HeLa cells (BOONE *et al.*, 1977).

Following the demonstration by WEI and MOSS (1974) of methylation of vaccinia virus mRNAs it was established that in addition to the 7'-methyltransferase activity, there occurs in the virion an S-adenosylmethionine:mRNA-(nucleoside -2'-0)-methyltransferase, which catalyzes reaction (d) in Fig. 9. This last enzyme has also been purified; it has a MW of 36K (BARBOSA and MOSS, 1978). In addition, an enzyme catalyzing conversion of 5'-phosphate and 5'-diphosphate termini of RNA to the triphosphate species has been characterized. It has a substrate specificity for ATP and partial specificity for dATP but not for the other nucleoside triphosphates (SPENCER et al., 1978).

In summary, mRNA synthesized *in vitro* by the RNA polymerase in the core of poxviruses becomes modified at the 3' end by the addition of covalently linked poly(A) and by being "capped" and methylated at the 5'-terminus. Specific enzymatic activities associated with modification of the mRNA have been identified. The significance of these modifications will be discussed in section V.

ii) Enzymes Related to DNA of the Virion

The occurrence of two deoxyribonuclease (DNAse) activities in the core of vaccinia virus, first established by POGO and DALES (1969a), was made more general for the poxviruses by the finding of similar activities in rabbitpox (AUBERTIN and McAUSLAN, 1972), Yaba virus (SCHWARTZ and DALES, 1971), and an insect poxvirus (POGO et al., 1971). This implies that such DNAses have a vital role in replicative functions. The two activities have pH optima of 4.5 and 7.8, respectively (POGO and DALES, 1969a; POGO and O'SHEA, 1977), and specifically utilize single-stranded (ss) DNA as substrate. The enzyme with the acidic pH optimum is an exonuclease, the other an endonuclease. Although both enzymes are active in the presence of the lateral bodies attached to the core (Fig. 5B), controlled proteolysis and coincident removal of the lateral bodies results in enhancement of both activities, particulary that of the endonuclease (POGO and DALES, 1969a). Following disruption of cores at high salt concentration ROSEMOND-HORNBEAK et al. (1974a, b) isolated an exonuclease activity having a pH optimum of 4.4, which also possessed a low level of endonuclease capacity. Although the exonuclease activity was contained in a polypeptide of MW 50K, existence of a dimeric structure of MW105K was considered by ROSEMOND-HORNBEAK et al. (1974a, b). A less drastic but more careful separation was conducted by POGO and O'SHEA (1977), whereby attention was paid to the residual activity of the endonuclease at each step of the purification scheme. Two independent activities identical with those occurring in the virion core could be separated by electrofocusing. These enzymes occur in two polypeptides with similar or identical MWs of 50K as determined by PAGE, but they differ in their surface charges so that they have different isoelectric points (PIs). It is conceivable that in the active state the DNAses exist as a heterodimer of MW ~100K, in accord with the suggestion of ROSEMOND-HORNBEAK et al. (1974b). One should mention that the presence of DNAse activities has been reported in other types of animal viruses (ROUGET et al., 1976), but as yet it is uncertain whether these are virus- or host cell-specified.

Suggestive evidence has been obtained concerning the functions of poxvirus core nucleases. By analogy with other eukaryotic systems one may postulate that they act to introduce nicks in the DNA template. The endonuclease is capable of specific removal of terminal cross-links between the complementary strands of the DNA genome (see section III). The cross-links are disrupted in an *in vitro* reaction with preparations of dissassembled cores (GESHELIN and BERNS, 1974), and also (apparently in the same manner) in host cell cytoplasm following penetration of the inoculum (POGO, 1977, 1980b). It has additionally been observed that, following inoculation with vaccinia virus, the DNAse-endonuclease activity appears in the soluble fraction of the cytoplasm, presumably having been released from the core, and it can be detected within the nucleus, where it may inhibit host DNA replication (POGO and DALES, 1973; POGO and DALES, 1974; OLGIATI *et al.*, 1976).

Modifications of the poxvirus genome, in preparation for either its replication or transcription, may also involve another core enzyme, the nicking-closing enzyme (BAUER *et al.*, 1977). This topoisomerase activity is capable of relaxing both left- and right-handed superhelical DNA. The enzyme has been solubilized and purified by chromatography on denatured DNA-cellulose columns. It appears to be a heterodimer complex consisting of two polypeptides of MW 35 K and 24 K. Of the two polypeptides, the 24 K species is basic in isoelectric point and more abundant, constituting 7% of the virion protein mass, but it is the much less abundant 35 K protein, which represents only 0.2% of the protein of the virion, which contains most of the nicking-closing activity. This raises the question as to whether the smaller, basic polypeptide is a *bona-fide* enzyme component. BAUER *et al.* (1977) offer the suggestion that association of the active 35 K subunit with the basic protein subunit may somehow be connected with regulation of the enzymatic interaction with the DNA. If the two combined polypeptides constitute the complete functional enzyme unit of MW ∼70 K, it was calculated that the average core should contain 125 complete enzyme molecules. The virion enzyme is distinguishable from similar host cell functions in the nucleus by differences in the ionic strength required for optimum activity and by size, both in terms of the s values obtained in velocity gradients and by MW determinations in PAGE. Although the biological function of the vaccinia virus nicking-closing enzyme is uncertain, it is suggested that the activity may have a role in transcription from the core, although an unknown function in DNA replication has also to be considered.

iii) Nucleoside Triphosphate Phosphohydrolase

The observed requirement during *in vitro* transcription for much greater amounts of ATP than those strictly required for the polymerization of mRNA and poly(A) led to the suggestion that ATP hydrolysis is obligatory for other processes, such as the extrusion of nascent mRNA chains from cores (KATES and BEESON, 1970a). Therefore, a rationale exists for the presence of a nucleoside triphosphate hydrolase(s) (ATPase) identified in vaccinia virions (GOLD and DALES, 1968; MUNYON *et al.*, 1968). This activity or activities could be localized both biochemically and by electron microscopic cytochemistry in the virus core (GOLD and DALES, 1968), as illustrated in Fig. 10. One should recall that a phosphatase activity was reported many years ago by SMADEL and HOAGLAND (1942) in purified vaccinia suspensions and this may, in fact, have been the ATPase now under

Table 5. *Structural or functional identity of some vaccinia virion polypeptides*

No.	Molecular weight 10^3 daltons	Core	Inter-mediate	Surface	Associated biological structure or function
1	~250	+			
2	~200	+			
3	158				
4	145				
5	138	+			RNA polymerase subunit (335)
6	131	+			RNA polymerase subunit (335)
7	120				
8	101				
9	97	+			guanine methyl transferase sub-unit (65)
10	89	+			polynucleotide 5′-phosphatase subunit (470)
11	86				
12	84.5	+			
13	79	+			RNA polymerase subunit (335)
14	76				
15	75			+	
16	72.5				
17	70	+			nucleotide phospholydrolase(s) I and II (353)
18	63.5	+			protein kinase
		+			nucleotide phosphohydrolase(s) I and II (353)
19	60	+			guanylytransferase with 7-methyl-transferase (296)
20	58			+	surface tubular elements (447)
21	57	+			poly A polymerase subunit (336)
22	55				
23	53.5				
24	51	+			poly A polymerase subunit ? (336)
25	50	+			ss endo- and ss exonucleases (193, 406)
26	48				
27	45				
28	43				
29	41.5		+		glycopeptide (148)
30	38				glycopeptide (148)
31	36.5	+			poly A polymerase subunit ? (336)
		+			RNA methylase (34, 483, 65, 311)
					DNA nicking-closing enzyme (39)
32	35	+		+	poly A polymerase subunit ? (331)
					RNA polymerase subunit
33	31	+			guanine methyl transferase sub-unit
34	29.5				
35	28.5			+	
36	27.5	+			
37	26.5	+			polynucleotide 5′-phosphatase subunit
38	25.5				
39	24.5				

Table 5. *Continuation*

No.	Molecular weight 10^3 daltons	Position in virion			Associated biological structure or function
		Core	Inter-mediate	Surface	
40	23				DNA nicking-closing enzyme component (39)
41	21.5				
42	20				
43	19.5				RNA polymerase subunit? (335)
44	~18	+			
45	~17	+			
46	~17	+			
47	16				RNA polymerase subunit? (335)
48	15				
49	~14			+	
50	~14				RNA polymerase subunit? (335)
51	13				
52	12				
53	~11				
54	~10.5	+			phosphorylated histone-like basic protein (392, 253)
55	~10	+			
56	<10			+	

The data in this table are compiled from information derived from slab PAGE, such as that illustrated in Fig. 7 A, the analysis of SAROV and JOKLIK (416) and other references indicated in brackets

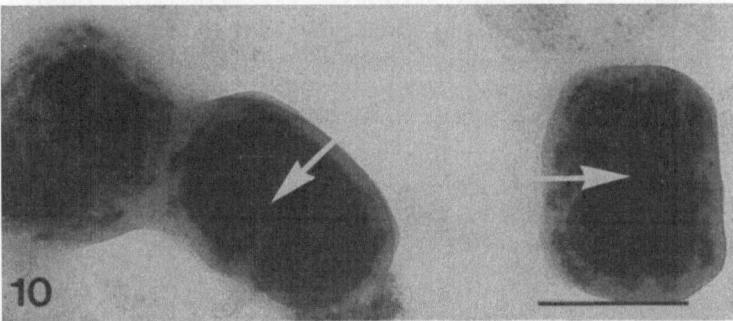

Fig. 10. Histochemical demonstration of nucleoside triphosphatase activity (NPHase) in whole particles of purified vaccinia virus. The reaction mixture contained ATP, $Pb(NO_3)_2$, and Mg^{2+} in Tris-maleate buffer. Following incubation at 37°, the dense lead phosphate product of the reaction (arrows) is deposited characteristically in the vicinity of the core but the surface membrane is free of lead phosphate deposits. × 120,000. Bar is 0.2 μm in length. From GOLD and DALES (1968)

consideration. Among the ribonucleotides tested, the ATPase has greatest affinity for ATP. However, the enzyme(s) appears to be even more active with deoxyribonucleotides (GOLD and DALES, 1968). In its solubilized state, the virion ATPase functions optimally at pH 8.0 in the presence of denatured DNA, and Mg⁺⁺ (PAOLETTI and Moss, 1972a). When cores are heated to 50° for 60 or 120 minutes the enzyme within them shows biphasic inactivation kinetics. In addition, ribonucleotides, especially ATP and UTP, protect the enzyme from heat inactivation (MUNYON et al., 1970). Following solubilization and partial purification on DNA-cellulose and Sephadex gels, two independent ATPase activities become apparent, ATPases I and II, presumed to exist as monomeric units of MW 61 K and 68 K, respectively. The ATPases are distinguishable on the basis of their heat lability and inactivation by cupric ions (WEI and MOSS, 1974; PAOLETTI and MOSS, 1974). It was calculated that on the average there are 100 to 300 ATPase molecules per virion (PAOLETTI et al., 1974). Although the biological functions of the ATPases have not been elucidated unequivocally, by analogy with other systems these enzymes have possible roles in DNA strand separation, initiation of DNA replication, transcription, extrusion of mRNAs from the core, DNA packing, virion assembly and the nicking-closing activity described above.

iv) Kinases

Among the kinases, the core-associated 5'-phosphate-polyribonucleotide kinase (SPENCER et al., 1978) has already been mentioned in connection with mRNA capping.

The presence among the virion polypeptides of phosphoprotein(s) (SAROV and JOKLIK, 1972a), focuses attention on the possible functions of such polypeptides and the mechanisms of their phosphorylation. Incubation of disrupted vaccinia virus particles or isolated cores under suitable conditions results in transfer of ^{32}P from γ^{32}-ATP to a virion acceptor protein (PAOLETTI and MOSS, 1972b; DOWNER et al., 1973; KLEIMAN and MOSS, 1973). The virion protein shown to be phosphorylated in the in vitro reaction is not, however, the major 11—12 K phosphoprotein labeled in vivo (SAROV and JOKLIK, 1972; POGO et al., 1975). A vaccinia virus protein kinase of MW 63 K has been solubilized and shown to consist of heat labile and heat stable components. Following addition of the latter component to the reaction mixture, the kinase activity is greatly enhanced. The enzyme has a dependence on Mg^{++} and can be stimulated by the presence of basic proteins in the reaction (KLEIMAN and MOSS, 1973). The time-course of in vivo phosphorylation of the histone-like 11—12 K basic protein suggests that phosphate groups are transferred by the protein kinase to threonine and serine residues in this and perhaps other less abundant basic polypeptides (POGO et al., 1975).

v) Alkaline Protease

A core activity of uncertain specificity and function was recently identified in vaccinia virions (ARZOGLOU et al., 1979). It remains to be established whether this alkaline protease is involved in post-translational cleavage related to the maturation of the virion.

H. Relationship of Virus Proteins and Functions

The complement of individual vaccinia virus polypeptides, as revealed by PAGE analysis, is illustrated in Fig. 7A, B. In Table 5, the number assigned to each band corresponds to the polypeptide identified by its estimated molecular weight, position within the virion and, whenever possible, association with a particular function(s).

III. Organization and Replication of the Genome

A. Organization

As might be anticipated from the complexity of their architecture and polypeptide content, the poxviruses contain a large molecular weight genome. By
direct visualization after spreading on protein films by the Kleinschmidt procedure
or by physical-chemical determinations, it is evident that the genomes exist as
double stranded linear DNA molecules about 45 to 60 μm long, (Fig. 11 A)
(RANDALL et al., 1966; HYDE et al., 1967; GAFFORD and RANDALL, 1967; GAFFORD
et al., 1978; ARIF, 1976; GESHELIN and BERNS, 1974). A presumed detection of
single stranded DNA within the virion (PFAU and McCREA, 1962 a, b) has not been
substantiated and this result could, conceivably, have been obtained artifactually
as a consequence of the extraction method employed, which produces fragments
of denatured DNA.

Table 6. *Physical-chemical properties of DNA genomes*

Virus type	Molecular weight in daltons × 10^6	Buoyant density (ρ)[a] in gm/cm^3	Unde-natured sedimen-tation value, S 20 w	References as listed in Bibliography	Year reported
Vaccinia, cowpox rabbitpox	160–170	1.691–1.695 (63–68)		200, 201, 337 195, 161	1962, 1965, 1967, 1968
Vaccinia, rabbitpox	118–130		61.4–72	72, 146, 196, 180, 161, 133, 53, 154, 135	1970, 1974, 1976, 1977
Avian e.g. fowlpox	200–240	1.695–1.689 (65)	78–82	145, 400, 185, 145 a, 324	1966, 1967, 1970
Avian, e.g. fowlpox	140–188		69–72	146	1977
Molluscum contagiosum	118			361	1977
Shope fibroma	153	1.700 (60)		195	1972
Entomopox	132–142	1.685 (73–75)		16	1976

[a] Values in brackets are the percentage content of A+T

Although the base composition and hence buoyant density (ρ) and MW values
vary from one poxvirus group to another, the range of values, shown in Table 6,
is relatively narrow (NISHIMURA, 1965). The fraction of A+T (Table 6) is higher
than the average content found in eukaryotic nuclear DNA, being closer to that
present in prokaryotes. In the case of parapoxviruses, not listed in Table 6, the
G+C fraction is 63% (WITTEK et al., 1979). The presence of minor bases in the
DNA has not been reported. Recent improvements in methods for estimating
MWs of DNA have necessitated the downward revision of the MWs of vaccinia
and other poxviruses, as shown by the data presented in Table 6. It has been

appreciated for some years that unless precautions are taken to avoid shear of the
intact genomes during preparation, fragmentation into molecules of about half
the normal length or less can occur readily (JOKLIK and BECKER, 1964; JAQUE-
MONT et al., 1972). Improved, new procedures for extracting molecules from virions
rapidly and in high yield have been developed (PARKHURST and HEIDELBERGER,
1976). These procedures utilize detergents, reducing agents such as mercaptoethanol and either high concentrations of urea (HOLOWCZAK, 1976) or concentrated
NaCl solutions to release the genome. The freed DNA may be concentrated into a
band in diatrizoate (iodinated salt) or other types of density gradient (ESPOSITO
et al., 1978), or it may be purified by ion exchange chromatography on hydroxy-
apatite columns (CABRERA and ESTEBAN, 1978).

Nucleic acid hybridization analyses have shown that the poxvirus genome
contains predominantly unique DNA sequences. However, by utilizing DNA-DNA
reassociation kinetics it became evident that 4 to 7% of the genome, representing
1.1 to 1.85×10^5 base pairs, consists of repeated sequences. Furthermore, among
the reiterated DNA segments, a sequence of 1.3×10^3 base pairs occurs as a multi-
copy repeat (PEDRALI-NOY and WEISSBACH, 1977; GRADY and PAOLETTI, 1977).
The terminal sequence is most probably the one in which numerous repetitions are
organized in tandem (WITTEK and MOSS, 1980, appended bibliography).

When isolated vaccinia virus DNA is denatured by sedimentation through
alkaline sucrose gradients, properties anomalous to linear dsDNA molecules
become apparent (BERNS and SILVERMAN, 1970; GESHELIN and BERNS, 1974). The
denatured molecules undergo rapid renaturation when returned to solutions at
neutral pH, implying that, despite rupture of hydrogen bonding, the sister strands
remain attached to each other. This attachment was shown to occur in the form
of two covalent nucleotide phosphodiester links, positioned at or near each
terminus of the genome (GESHELIN and BERNS, 1974). Genomes of poxviruses other
than vaccinia virus were also shown to possess similar interstrand links (GAFFORD
et al., 1978; JAUREGUIBERRY, 1977; McCARRON et al., 1978; WITTEK et al., 1978a;
PARR et al., 1977; DEFILIPPES, 1976). It has recently been demonstrated that DNA
crosslinking may be a more general feature, as suggested by observations made on
chromosomal DNA from yeast nuclei (FORTE and FANGMAN, 1976). Denatured
vaccinia virus DNA observed by electron microscopy of spread molecules on
protein films is circular (GESHELIN and BERNS, 1974; HOLOWCZAK, 1976), as
shown in Fig. 11B, confirming that the separated DNA strands are physically
joined near the ends of the molecule. Evidence for the phosphodiester nature of
the cross-links was obtained by exposing the native viral genomes to controlled
hydrolysis by the virion-associated endo-DNAse, which brought about their
elimination (GESHELIN and BERNS, 1974). The hypothesis that the cross-links
may have a function in circularization of the genome for the purpose of replication
of the DNA has been suggested on the basis of experimental data (POGO, 1977;
ESTEBAN et al., 1977; McFADDEN and DALES, 1979). Termini of the genome
containing the cross-links have been isolated by means of hydrolysis with restriction endonucleases, followed by electrophoresis through agarose to separate the
discrete DNA fragments (JAUREGUIBERRY, 1977; McCARRON et al., 1978; WITTEK
et al., 1977). More detailed information concerning the nature of crosslinks, their
position in the genome and function in DNA replication can be anticipated in the

near future. It has already been shown that the two DNA fragments of the opposite termini in the rabbitpox genome contain sequence homology (WITTEK *et al.*, 1977).

Thus, two technological developments, rapid isolation in high yield of pure, intact poxvirus genomes and the use of a large variety of bacterial restriction endonucleases to generate specific cleavage fragments of DNA, presage the accumulation of much information providing insights into the nature of the poxvirus genome. By employing the endonucleases, it is now feasible to construct "restriction" maps on which an orderly positioning of specific mRNA functions can be made, as already evident in the work of WITTEK *et al.* (1977, 1980 b) and CABRERA

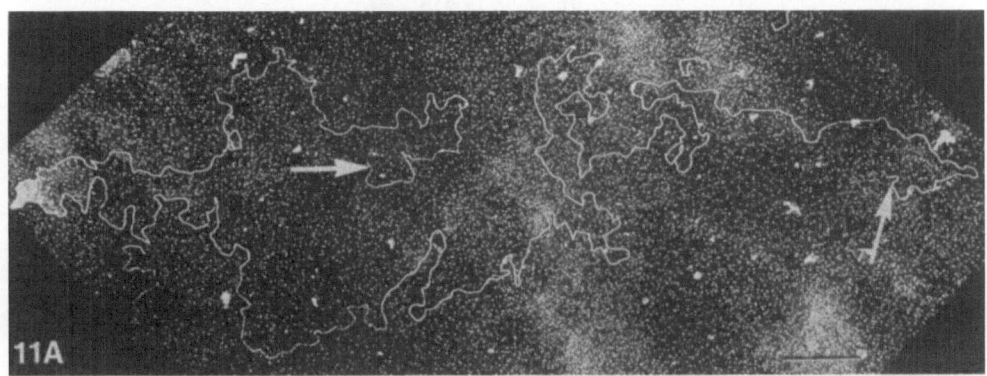

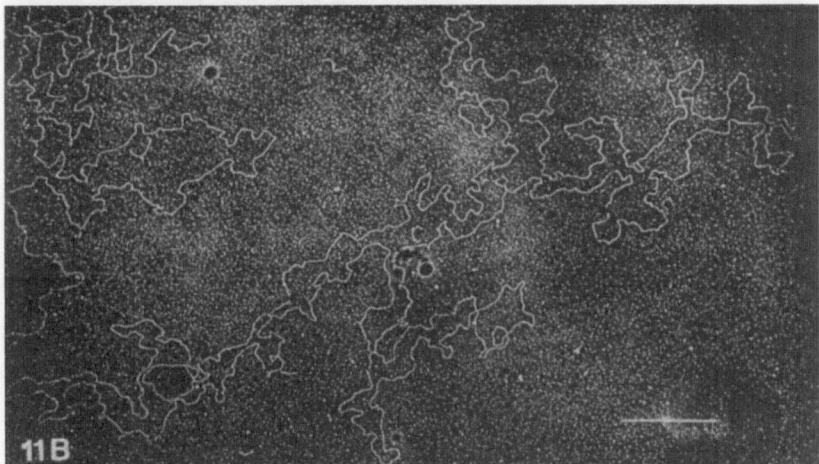

Fig. 11. *A* A single molecule of vaccinia virus DNA genome spread on a thin film of protein and stained with uranyl acetate. The distance between the free ends of this molecule (arrows) is equivalent to a molecular weight of approximately 120×10^6. The image was printed in reverse to enhance contrast ($\times 13{,}000$). (K. L. EGGEN and S. DALES, unbublished)

B A similar preparation to that in *A*, illustrating a vaccinia virus DNA molecule following controlled denaturation at elevated temperature. Note that the thread of ssDNA is finer and more flexible than that shown in panel A and the molecule is in a closed, circular configuration. $\times 16{,}000$. (From GESHELIN and BERNS, 1974.) Bars are 1 μm in length

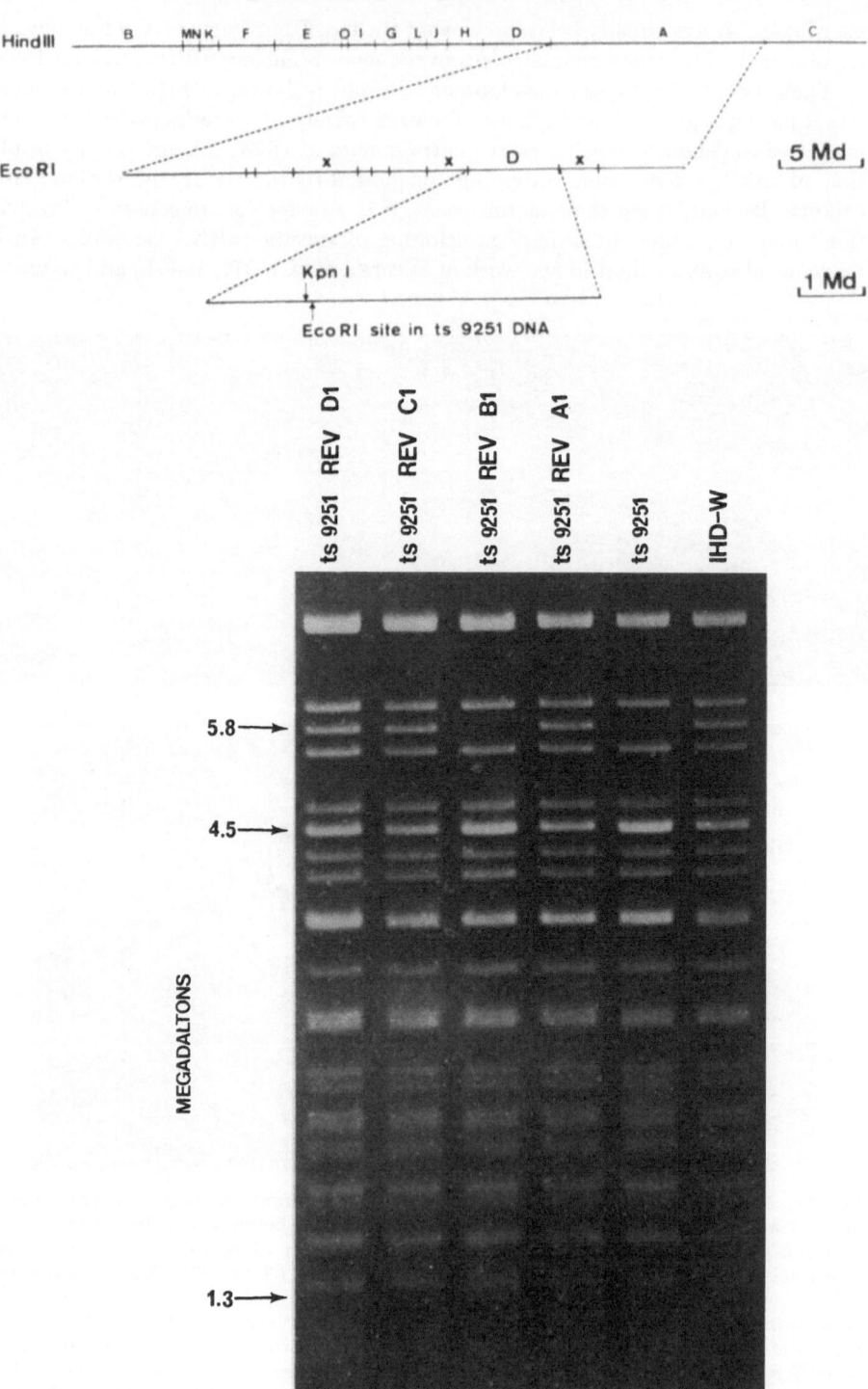

Fig. 12

and colleagues (1978). Use of endonucleases allows the determination of genetic relatedness between various groups of ortho- and parapoxviruses (ARCHARD and MACKETT, 1979; WITTEK *et al.*, 1977; ESPOSITO *et al.*, 1978; WITTEK *et al.*, 1980a), and can also reveal the most minute genetic change within a particular locus on the map (Fig. 12) (McFADDEN *et al.*, 1980; SCHÜMPERLI *et al.*, 1980). A comparison of the homology among the internally positioned endonuclease fragments from ortho- and parapoxviruses has already shown extensive homology in the informational content among the agents within each group. Further analysis of cross-linked or "snap-back" terminal fragments with respect to their base sequence reveals that there has been an evolutionary conservation among orthopoxviruses in this portion of the virus genome (WITTEK *et al.*, 1978a, b; MÜLLER *et al.*, 1977; WITTEK *et al.*, 1977; MENNA and WYLER, 1977; GARON *et al.*, 1978; MENNA *et al.*, 1979).

Genomes of DNA viruses, as they exist within virus cores, are complexed with specific proteins of host or virus derivation. This has been shown with the papovavirus group, as well as with adenoviruses and herpesviruses. In the case of the poxviruses all polypeptides identified as a complex with the DNA, whether obtained from virions or from cytoplasmic "fractions", appear to be virus-specified. The question, however, remains as to which among these polypeptides are specifically attached to the DNA and function in the manner of chromatin-associated proteins. Nucleoprotein complexes obtained from the cytoplasm of cells in which vaccinia virus replication is in progress contain either several (POLISKY and KATES, 1976), or many (SOLOSKI *et al.*, 1978; McFADDEN and DALES, 1980) DNA-binding proteins of heterogeneous MWs, ranging from 12.5 to 90 K. Some of these proteins must be complexed to the DNA via ionic bonds, as evident by the ability to dissociate them in a reversible manner by the addition or removal of 2 M NaCl (POLISKY and KATES, 1976). In the nucleoprotein complexes may be included the acidic 28 K polypeptide and the phosphorylated, basic and lysine-rich 34 K polypeptides (NOWAKOWSKI *et al.*, 1978), which might be constituents of the DNA topoisomerase or nicking-closing enzymatic activity identified in vaccinia cores (BAUER *et al.*, 1977). A very basic, phosphorylated 10—11 K polypeptide, presumably identical with the one present in large amounts within virion cores

Fig. 12. Agarose gel electrophoresis of EcoRI-digested DNA from a wild-type vaccinia virus mutant, *ts* 9251, and four spontaneous *ts+* revertants of 9251. Virus stocks were prepared, viral DNA was extracted on lysis gradients, and the purified DNA was digested with EcoRI endonuclease as described by McFADDEN and DALES (1979). Electrophoresis was performed for 16 hours at 50 V in 0.7% agarose and the DNA was visualized with 1 μg/ml ethidium bromide. Molecular weight markers were the products of λ plac 5 digested with EcoRI or HindIII. In the diagram above is drawn the map position of the newly acquired EcoRI restriction site of *ts* 9251. Top: HindIII restriction map of rabbit poxvirus DNA drawn according to WITTEK *et al.* (1977). In vaccinia IHD-W, fragments B and C are reversed. Middle: EcoRI restriction map of HindIII fragment A. Bottom: Enlarged representation of EcoRI fragment D of rabbit poxvirus DNA or vaccinia IHD-W denoting the novel EcoRI site of *ts* 9251. Segments marked (×) have not been mapped precisely to date. KpnI endonuclease site in fragment D is also indicated

(POGO *et al.*, 1975), is also found in the cytoplasmic complexes. This histone-like material has been postulated to bind preferentially to superhelically coiled DNA (NOWAKOWSKI *et al.*, 1978). Although both spermine and spermidine have been shown to occur in vaccinia virus cores (LANZER and HOLOWCZAK, 1975), it remains uncertain whether either of these two polyamines is directly bound to the DNA.

Based on autoradiographic and biochemical evidence it has long been recognized that replication of poxvirus DNA occurs in the cytoplasmic matrix (HARFORD *et al.*, 1966; KATO *et al.*, 1963b, 1962b; KAJIOKA *et al.*, 1964; MAGEE and SAGIK, 1959; MAGEE *et al.*, 1960; SHEEK and MAGEE, 1961; CAIRNS, 1960). The cytoplasmic factories where the DNA is synthesized were shown to be the foci in which uncoated parental genomes are concentrated (DALES, 1963). Although it is the generally accepted view that host DNA is not involved in virus DNA replication, it was suggested that infection of HeLa cells causes hydrolysis and subsequent reutilization of host DNA as a substrate for vaccinia DNA synthesis (OKI *et al.*, 1971). This idea could not, however, be substantiated by later work from the same laboratory (PARKHURST *et al.*, 1973). Likewise the notion that vaccinia virus DNA replication is not confined exclusively to the cytoplasmic compartment (LA COLLA and WEISSBACH, 1975) and somehow involves the nucleus has been revised recently (BOLDEN *et al.*, 1979), thereby minimizing even further the possibility that the nucleus is an alternative site of poxvirus DNA synthesis. On the contrary, there is good evidence that the host-cell nucleus is most probably not involved. A case in point is the capability of vaccinia virus DNA to replicate in the cytoplasm of cells pretreated with the antibiotic mitomycin C, which modifies host DNA by crosslinking and abrogates its template functions (KAJIOKA *et al.*, 1974; MAGEE and MILLER, 1962). Furthermore, the ability of cytoplasts, created by means of enucleation with cytochalasin B, to synthesize functional vaccinia virus DNA after infection is additional proof that the poxvirus genome exercises autonomy in its own replication (PENNINGTON and FOLLET, 1974; PRESCOTT *et al.*, 1971).

Studies in several laboratories have demonstrated that among the poxviruses the process of genome duplication follows the semiconservative, symmetrical pattern (POGO and O'SHEA, 1978; HOLOWCZAK and DIAMOND, 1976; LAMBERT and MAGEE, 1977; ESTEBAN and HOLOWCZAK, 1977a, b, c). In systems that are amenable to synchronous infection, biochemical and autoradiographic analyses reveal that the bulk of the DNA is synthesized in a wave of relatively short duration. For example, in cultured cells infected by vaccinia virus, the duration of DNA synthesis is only about 120 minutes, commencing at 1.5 hours and subsiding by 3.5 to 4 hours post inoculation (JOKLIK and BECKER, 1964; SHEEK and MAGEE, 1961; Cairns, 1960). When infection by more slowly multiplying agents such as Yaba and rabbit fibroma viruses is analysed, the intervals during which DNA is produced may extend for 12 hours or longer (YOHN *et al.*, 1970; EWTON and HODES, 1967). Inoculation of defined tissues, such as the skin or scalp of chickens, with fowlpox virus elicits enhanced *in vivo* DNA synthesis, some of which may initially be host-related, but which later becomes virus specified, continuing for 24 hours or longer (CHEEVERS and RANDALL, 1968).

With all poxviruses examined, there is an obligatory requirement for protein synthesis preceding the onset of DNA replication. This is evident even after the uncoating stage has been passed (JOKLIK and BECKER, 1964), suggesting that enzymatic activities required for DNA formation must be acquired. Among these enzymes which appear in the cytoplasm of infected cells are a DNA-dependent DNA polymerase, deoxyribonucleotide kinase(s) (BERNS et al., 1969; CHALBERG and ENGLUND, 1979; JOKLIK, 1962a, JUNGWIRTH and JOKLIK, 1965; GREEN and PINA, 1962; MAGEE, 1962; MAGEE and MILLER, 1967; ERON and McAUSLAN, 1966; CITARELLA et al., 1972; BERGER et al., 1978), and the DNAse which acts on double stranded DNA at an alkaline pH (McAUSLAN, 1965). The DNA polymerase becomes manifest coincidentally with commencement of virus DNA formation, rises to a maximum by 4 hours and retains high activity for several more hours, even after shut-off of DNA synthesis (JUNGWIRTH and JOKLIK, 1965; GREEN and PINA, 1962). Identification of the cytoplasmic DNA polymerase as a virus-specified function has been established by several criteria, including differences from host DNA polymerases with respect to responses to primers, activity at elevated pH, thermolability, requirements for divalent cations, inactivation by enzyme poisons and by specific antibody (BERNS et al., 1969; MAGEE and MILLER, 1967). The virus-related DNA polymerase has been recovered from the cytoplasm of infected cells and purified to homogeneity (CHALBERG and ENGLUND, 1979). The activity occurs in a single polypeptide of MW 110—115K, and is intimately linked to an exoDNAse (CHALBERG and ENGLUND, 1979), implying an essential function for this nuclease during duplication of the genome.

Synthesis of stoichiometric amounts of other virus-specified proteins related to the DNA, including a DNA polymerase and DNA ligase (MAGEE, 1962; SAMBROOK and SHATKIN, 1969), appears to be required throughout the period of genome duplication (KATES and McAUSLAN, 1967b; BEDSON, 1968). One of these proteins participates in the insertion of covalent cross-links at genome termini (vide infra, ESTEBAN and HOLOWCZAK, 1978; POGO, 1979, 1980a).

While synthesis of poxvirus DNA conforms in many respects to semiconservative replication patterns established for entities ranging from the bacteriophages to eukaryotic cells, unique features peculiar to the duplication of poxvirus genomes have also been recognized. The most notable among these features is the presence of the aforementioned terminal cross-links (BERNS and SILVERMAN, 1979). Existence of such interstrand connections presumably imposes constraints on genome replication, necessitating their removal to permit initiation of the strand copying process. It has, in fact, been observed that following uncoating of the parental genome in the cytoplasmic matrix, the molecule undergoes rapid modification by elimination of the cross-links. This change becomes evident from sedimentation analysis in alkaline and neutral sucrose gradients which can reveal formation of genome-length single-stranded DNA (POGO, 1977). Removal of cross-links can also be accomplished by exposing intact vaccinia virus genomes either to cytosol from recently inoculated cells (POGO, 1977), or to ss endoDNAse partially purified (GESHELIN et al., 1974) from vaccinia virus cores (POGO and DALES, 1969a, POGO and O'SHEA, 1977). Taken together, these observations offer evidence for the role of the virion endonuclease in modification of the genome before replication can begin. The cross-links are reinserted into progeny genomes during terminal stages of genome formation (HOLOWCZAK and DIAMOND, 1976; POGO and O'SHEA, 1978; POGO, 1979).

B. Replication

Replication of the genome is a multi-step process involving initially the synthesis of small 10 to 12S single-stranded DNA fragments which are covalently linked with RNA primer sequences (POGO and O'SHEA, 1978; HOLOWCZAK and DIAMOND, 1976). The short segments become assembled into an intermediate double-stranded DNA structure of 70S unit length (POGO and O'SHEA, 1978) and finally the genome molecules. Concatameric structures or "rolling circles" such as exist with other viruses, have not been identified as intermediates in the poxvirus replication process and probably do not exist *in vivo* under normal conditions.

A recent model for vaccinia virus DNA synthesis has been proposed (ESTEBAN *et al.*, 1977), which is derived from an assumption that duplication and elongation are initiated at one end of a circular DNA structure. However, the existence of inverted terminal repeat sequences makes a model for initiation at a particular end less likely. This idea is further discounted by the discovery of spontaneously occurring non-lethal, terminal, mirror-image deletions (McFADDEN and DALES, 1979), indicating that genome replication most probably involves a circular structure. Evidence for circularity of replicating vaccinia virus DNA is also provided by sedimentation data employing alkaline sucrose gradients. When genomes isolated at the time of duplication are analysed by this means, it becomes evident that in addition to the mature, cross-linked DNA strands there exist molecules which have properties of open-circular structures as determined by the rate of sedimentation and partial binding of ethidium bromide (ARCHARD, 1979). Such partially circular molecules contain all sequences encoded in the virus genome. In a recent study, replicating molecules were annealed to isolated strands of vaccinia virus DNA and after hybridization the double stranded DNA thus created was subjected to restriction endonucleases (this procedure makes it possible to ascertain the specific activity of nascent DNA present in the annealed fragments). The data obtained indicate that initiation and termination of DNA synthesis occur bidirectionally, commencing at both ends of the genome by a process similar to that described for adenovirus DNA replication. The possibility that the displaced parental strand can undergo circularization has also been considered (POGO and O'SHEA, 1979).

Vaccinia virus DNA replication has also recently been achieved in an *in vitro* system. Cytoplasmic fractions containing factories from infected cells are able to catalyze the incorporation of deoxyribonucleotides into DNA in a reaction which is dependent upon the presence of all four deoxyribonucleoside triphosphates and is enhanced by addition of ribonucleoside triphosphates, suggesting that the latter are involved in a priming reaction (LAMBERT and MAGEE, 1977; ESTEBAN and HOLOWCZAK, 1977b). The product appears in the form of short 10S chains which, in time, become elongated to ∼70S DNA (LAMBERT and MAGEE, 1977; ESTEBAN and HOLOWCZAK, 1977a, b, c). The virus DNA polymerase involved has been isolated from such cytoplasmic fractions, partially purified and characterized (MAGEE and MILLER, 1967; BERNS *et al.*, 1969; CITARELLA *et al.*, 1972; CHALBERG and ENGLUND, 1979). The presence of an associated exonuclease activity was also reported in the most recent studies cited.

IV. Events During Penetration

A. Adsorption

The efficiency of poxvirus attachment to the host cell, as with any virus, depends directly on its collision frequency with the cellular surface. The probability of a collision increases as virion concentration goes up or it may be enhanced by the application of centrifugal force (SHARP and SMITH, 1960). However, the observed rates of attachment fail to reach values predicted from theoretical calculations (SMITH and SHARP, 1960; DUMBELL et al., 1957; ALLISON and VALENTINE, 1960a). Perhaps this is a consequence of an instability in the initial cell-virus association, which appears to involve electrostatic forces between the predominantly acidic groups on the host and basic residues on the virus (ALLISON and VALENTINE, 1960b). While this notion that binding involves primarily electrostatic forces might be an oversimplification, at least it accounts in part for the ability of vaccinia virus and other poxviruses to become adsorbed to a broad spectrum of cell types and suitably charged non-biological materials, among them glass and metallic surfaces. Very little information is presently available concerning the next stage, when specific, irreversible binding between virus ligands and host receptors occurs, although receptors for vaccinia virus and related serotypes are ubiquitous among a great variety of mammalian and avian cells, bringing into question the nature and specificity of such receptors. Agents of the parapox group likewise demonstrate tropism for a wide range of mammalian species, as exemplified by viruses from goats and sheep such as Orf virus which can replicate in cells of the human epidermis or in culture (ANDREWES and PEREIRA, 1972). By contrast, *Molluscum contagiosum*, a human orthopoxvirus unrelated to vaccinia virus, is avidly adsorbed *in vitro* to freshly explanted or established lines of primate cells and to freshly explanted embryonic avian and murine fibroblasts (ROBINSON et al., 1969; POSTLETHWAITE and WATT, 1967), but is attached inefficiently to transformed mouse fibroblasts (McFADDEN et al., 1979). These observations imply that specific receptors for this virus may generally exist on cells of primate species and species of lower homoiotherms while these cells are in an embryonic state. Sometimes receptors occur on specialized regions of the plasma membrane, such as those for insect poxviruses located on microvilli of intestinal epithelial cells lining the gut (Fig. 13, GRANADOS, 1973b).

The virion component involved in specific attachment to the cell surface has not been identified, but at least in the case of vaccinia virus, it may be connected with the STE (Fig. 2), as suggested by data showing that monospecific antibody to STE also possesses virus neutralizing activity (STERN and DALES, 1976b). It must, however, be pointed out that, unlike the situation with viruses like influenza virus and polio virus (DALES, 1973), which cannot be adsorbed by cells after complexing with antibody, antibody-neutralized vaccinia virus does become attached to cells, albeit at 30 to 50% reduced efficiently, perhaps because some adsorption sites on the very large virus surface escape binding of antibody. Since virions enclosed in host-derived wrapping membranes can also become adsorbed to cells and initiate infectivity, effective attachment is not limited to interactions with the envelope of the naked particle (PAYNE and NORRBY, 1978).

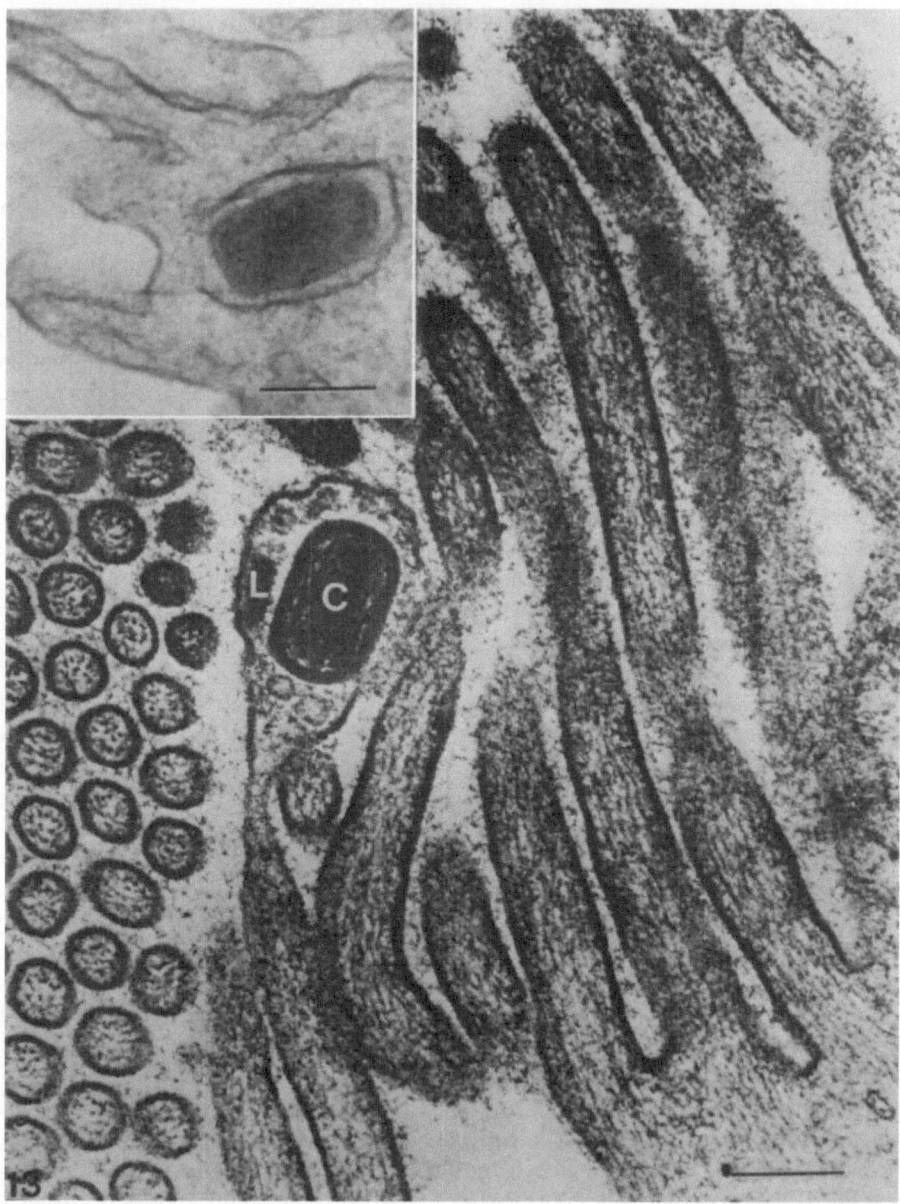

Fig. 13. Penetration of *Amsacta moorei* pox into a microvillus of an intestinal epithelium cell of an *Estigmene acrea* larva. Fusion between the virion and plasma membrane must have occurred prior to release of the core *(C)* and separation of a lateral body *(L)*. (Photomicrograph courtesy of R. R. GRANADOS.) Inset illustrates phagocytosis of vaccinia virus by a murine L cell fibroblast. × 85,000. Bar is 0.2 μm in length

When it is desired to synchronize the infectious process in work with *in vitro* systems, the quantity of inoculum added must be adequate to infect all cells in a culture. Although the concentration of infectious virus can be determined accurately by the pock or plaque assay and by other standard procedures, the concentration of infectious units does not provide a measure of the total virions participating. As discussed earlier, ratios of physical to infectious units can vary widely depending on the source of the virus, procedures used for isolation and purification, as well as susceptibility of a given host cell (OVERMAN and TAMM, 1957; DUMBELL *et al.*, 1957). In the usual experiments involving the standard *in vitro* one-step growth cycle, adsorption is allowed to take place at 0° to 4° when virus added at a multiplicity of 3 to 100 PFU per cell (representing 10 to 10,000 virions) is continuously mixed with 10^6 to 10^7 cells per ml in either stationary or suspension cultures. If the attachment occurs efficiently, over 80% of the inoculum can be adsorbed within 30 minutes and more than 90% within 60 minutes (DALES, 1963), although lower rates have also been measured (SMITH and SHARP, 1960). Following adsorption at low temperature, penetration of the cell membrane by the inoculum is initiated by warming the cultures to 37° C (DALES, 1973). Correlations have been established between the PFU's adsorbed and virus particles attached at the cell surface by comparing disappearance of inoculum PFU's from suspension with the number of attached inoculum virions enumerated by electron microscopy (DALES, 1963).

B. Penetration from the Surface

During penetration, the structure of poxviruses mandates that, as the membrane barrier is traversed, the virion envelope is removed so as to create a passage into the cytoplasmic matrix for the genome-carrying virus core. Passage through the plasma membrane is achieved at the point of cell-virus contact, which becomes the site of envelope fusion. Fusion may occur either externally, at the surface, or following engulfment of the inoculum inside a vacuole formed by invagination of the plasma membrane. This latter process is termed viropexis. There is, however, no fundamental difference between the two routes of penetration, since both require fusion between the lipoprotein barriers of the host and virus (Figs. 14 A—C). Electron microscopic data from both *in vivo* and *in vitro* experiments designed to trace the fate of inoculum virions reveal that viropexis is the more common route of entry into mammalian and avian cells (DALES and SIMINOVITCH, 1961; DALES, 1963; DALES and KAJIOKA, 1964; ROBINSON *et al.*, 1969; SHAND *et al.*, 1976; VREESWIJK *et al.*, 1976, 1977; PROSE *et al.*, 1969). However, in the case of vaccinia virus, at least, fusion at the surface clearly can occur (DALES, 1973; ARMSTRONG *et al.*, 1973; DALES *et al.*, 1976). Studies on insect poxviruses infecting their larval hosts reveal that penetration occurs by surface fusion at microvilli of intestinal epithelial cells (GRANADOS, 1973), whereas viropexis is the preferred pathway when circulating hemocytes are infected (DUVAUCHELLE *et al.*, 1971).

Under some circumstances, progeny virions are released from host cells in an orderly process involving wrapping by Golgi membranes (DALES, 1963). When such virions possessing the extra shroud constitute the inoculum for *in vitro* studies of penetration, it has been reported that penetration is more efficient than

with naked virions (PAYNE and NORRBY, 1978). However, the mechanism of this
mode of penetration has not been elucidated.

Following fusion of virus with the cell plasma membrane, antigenic components
of the vaccinia virus envelope are rapidly and widely dispersed within the plane
of the cell membrane, as revealed by immunoferritin tagging (CHANG and METZ,
1976). The ultimate fate of these antigens is obscure. However, in the case of
phospholipids of the virion inoculum it has been shown that $^{32}PO_4$ in the phospholipids is rapidly and efficiently solubilized and released to the extracellular
milieu (JOKLIK, 1964a).

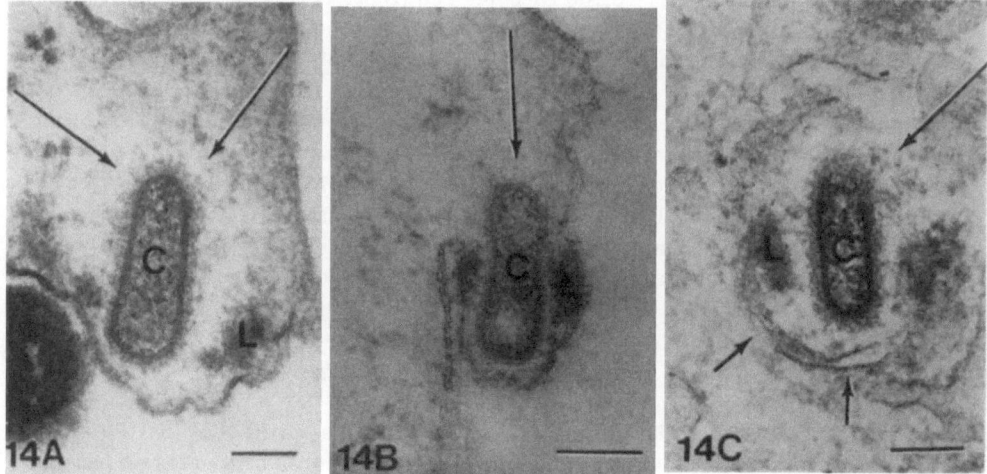

Fig. 14. Three examples of vaccinia virus penetration into host cells observed by means
of thin sections. *A* Stationary cultures of HeLa cells were inoculated, sampled 10 min
later, and prepared for electron microscopy without disturbing the *in vivo* relationships.
An extracellular virion and a fused particle are evident in this example. *B* and *C* Strain
L cells in suspension were sampled 20 min after initiating the infection. The long
arrows indicate a region of lysis whereby the core(s) can be moved into the cytoplasmic
matrix. The short arrows in *A* point toward membrane segments of the phagocytic
vacuole and virus envelope remaining deep in the cytoplasm after fusion and lysis
have occurred. *B* The virion envelope became fused with the cell membrane before
phagocytosis could occur. It should be noted that in each case the lateral bodies *(L)*
are separated from the core *(C)* and remain attached to the envelope. *A* × 95,000;
   B × 128,000; *C* × 107,000. (From DALES *et al.*, 1976.) Bars are 0.1 μm in length

The simultaneous fusion of virus and cell membranes is undoubtedly under the
control of an external component on the virion, perhaps associated with the STE.
This idea stems from the finding that monospecific anti-STE antibody interferes
with virus penetration (STERN and DALES, 1976b). The fusion-controlling factor
is heat labile, because heating for 30 minutes to 56—60° also suppresses the fusion
event (DALES and KAJIOKA, 1964). Both heat-denatured and antibody-neutralized
vaccinia virus can adsorb to cells and become phagocytized, but in each case the
virus fails to enter the cytoplasm from the enclosing vacuole by the normal route.
Instead, the inactive virus is carried within vacuoles directly into lysosomes, where

complete dismemberment and hydrolysis of its protein and DNA occur (Fig. 15 and Table 7). This heterophagic process within the "garbage-disposal" lysosomal organelles should be viewed as a manifestation of a generalized defense mechanism for eliminating infectious agents. In the normal infectious process with poxviruses, this mechanism does not operate, because the undenatured inoculum is routed through the cell membrane by means of lysis directly into the cytoplasmic matrix.

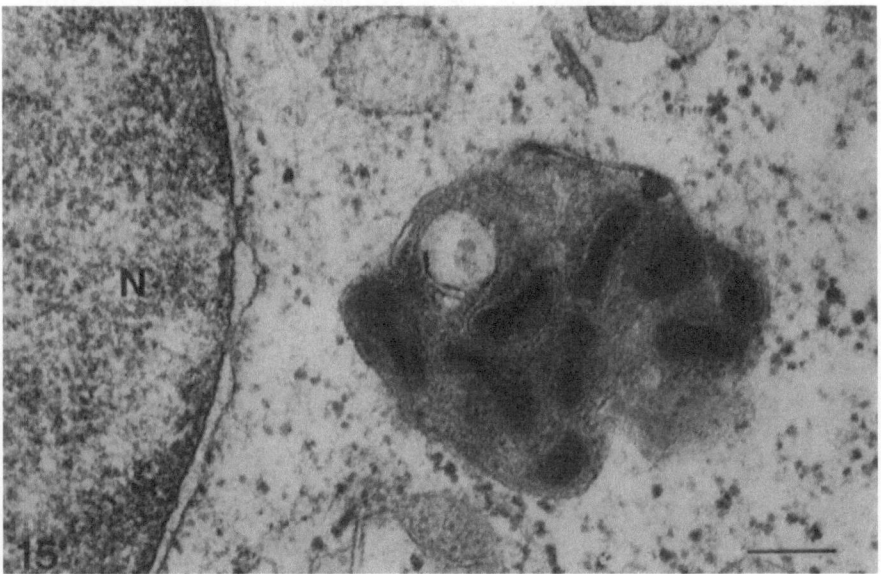

Fig. 15. Selected area of an L strain mouse fibroblast illustrates the nucleus and cytoplasm containing a lysosome filled with tightly packed inoculum vaccinia virus particles. Heat-denatured or antibody-neutralized inoculum particles fail to penetrate by simultaneous lysis of the cell membrane with the virus envelope, but instead are carried into lysosomes where the virions become degraded. × 60,000. (From DALES, 1969)

C. The Phenomenon of Uncoating

The uncoating process represents the terminal stage in the release of the viral genome from its protecting core. This process can be visualized by electron microscopy (Figs. 16 A—C) and monitored biochemically, in terms of the quantity of DNA in the inoculum which becomes accessible to DNAse hydrolysis. It might be surmised that agents as complex as poxviruses would exercise some direct control over the uncoating process. The first evidence for the notion that the incoming virion somehow triggers its own replicative process in the cytoplasm originated from the analysis of CAIRNS (1960). He pointed out clearly that when a particular host cell has been permeated by several infectious vaccinia particles, each with a capacity to establish an independent cytoplasmic center or factory for DNA synthesis, the multiple factories are turned on simultaneously, as if by some critical, initiating event. In fact, the lag period to DNA synthesis can be curtailed by increasing the inoculum multiplicity and the lag is virtually abolished

Table 7. *Counts of autoradiographic grains over sections of L cells sampled at intervals following inoculation with H³-thymidine-labeled vaccinia virus*

Hours post-infection	Control					Antiserum-treated				
	Grains over nucleus	Actual grains in cytoplasm	Per cent label conserved	Virus particles per cell profile Total	Intracellular virus particles per cell profile (Total minus surface)	Grains over nucleus	Actual grains in cytoplasm	Per cent label conserved	Virus particles per cell profile Total	Intracellular virus particles per cell profile (Total minus surface)
1	0.12	2.93	100	1.25	1.02	0.04	1.70	100	0.90	0.72
4	0.16	2.87	98	1.00	0.93	0.18	1.08	64	0.46	0.37
8	0.10	2.16	73			0.18	0.63	37	0.27	0.19
1	0.18	10.92	100	3.90	2.94	0	3.58	100	2.76	1.69
3	0.15	10.56	97	2.54	2.30	0	2.26	63	1.74	1.27
6	0.04	8.55	78			0.01	1.72	48	0.93	0.67

(DALES and KAJIOKA, 1964.)

for a superinfecting inoculum that is added after the time interval normally occupied by the lag period (JOKLIK, 1964). Evidence for a spatial-temporal relationship between the activation of virus synthetic foci and uncoating was obtained by tracing the fate of ^{3}H-thymidine labelled inoculum from the surface into the cytoplasm by means of light and electron microscopy combined with autoradiography (DALES, 1963). These studies revealed that the release of DNA occurs synchronously from a multiplicity of cores present in any given cell inoculated with an average multiplicity of 10 or more infectious units per cell. This finding may be related to the critical event required for switching on the factories (CAIRNS, 1960). The viral DNA, presumably complexed with a specific set of internal proteins, passes through clear-cut breaks in the proteinaceous core coat

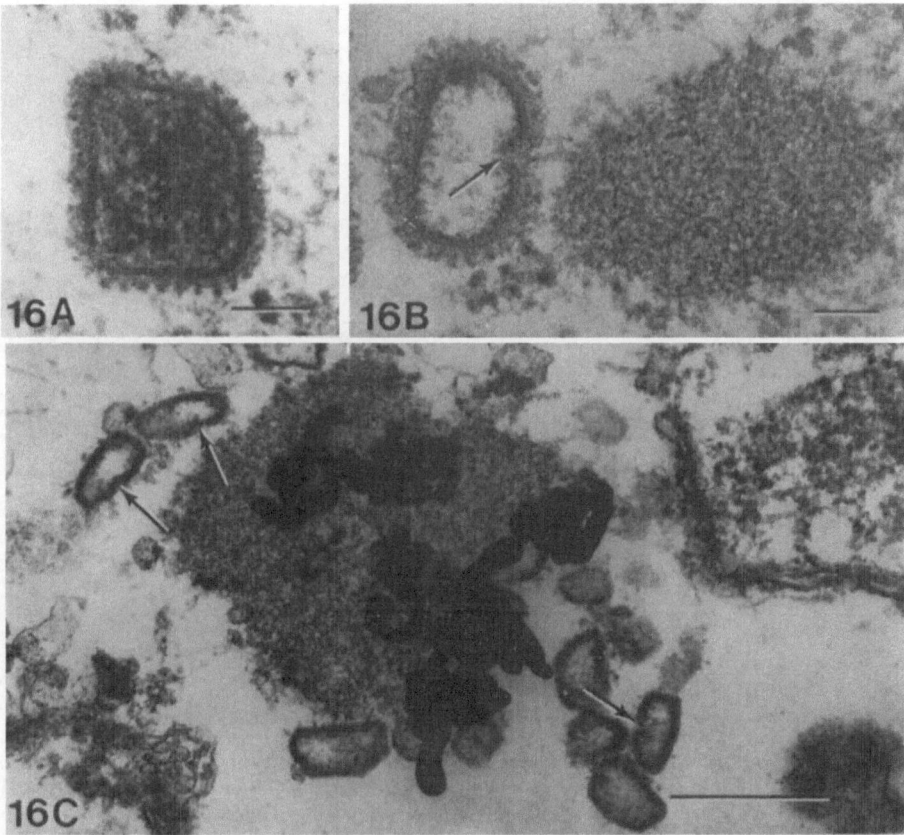

Fig. 16. Intracytoplasmic uncoating of vaccinia virus cores. *A* the thick membrane of the dense core is covered by fine projections. *B* and *C* release of material from cores into the cytoplasmic matrix. Puddles of dense fibrous material accumulate in the vicinity of shells (i.e. remnants of cores), frequently on the side facing a break in the thick membrane (arrows). The autoradiogram *(C)* reveals that DNA passes out of cores during uncoating and is concentrated in the fibrous matrices. Single cores or groups are uncoated simultaneously in any one cell, presumably after synthesis of the uncoating factor. *A* and *B*, ×110,000; *C*, ×90,000. (*A* from DALES and KAJIOKA, 1964; *B* from DALES, 1965a; *C* from DALES, 1965b)

(DALES, 1965a), leaving behind the empty cores, termed *shells* (Figs. 16 B, C) (DALES and KAJIOKA, 1964). Under circumstances involving cooperative uncoating of several or many cores in a restricted region of the cytoplasm, it is found that the breaks develop characteristically on the side facing the pools of discharged genome material (Fig. 16 B, C) (DALES, 1964b), as if rupture occurs as a specific response generated by adjacent virions. Following uncoating, the labelled DNA from the inoculum is conserved within the "factories" throughout the replicative cycle (Fig. 17, Table 7), and occasionally it may even pass into progeny virions (DALES, 1965a). It is, however, not known whether the conserved DNA is all or only part of each genome.

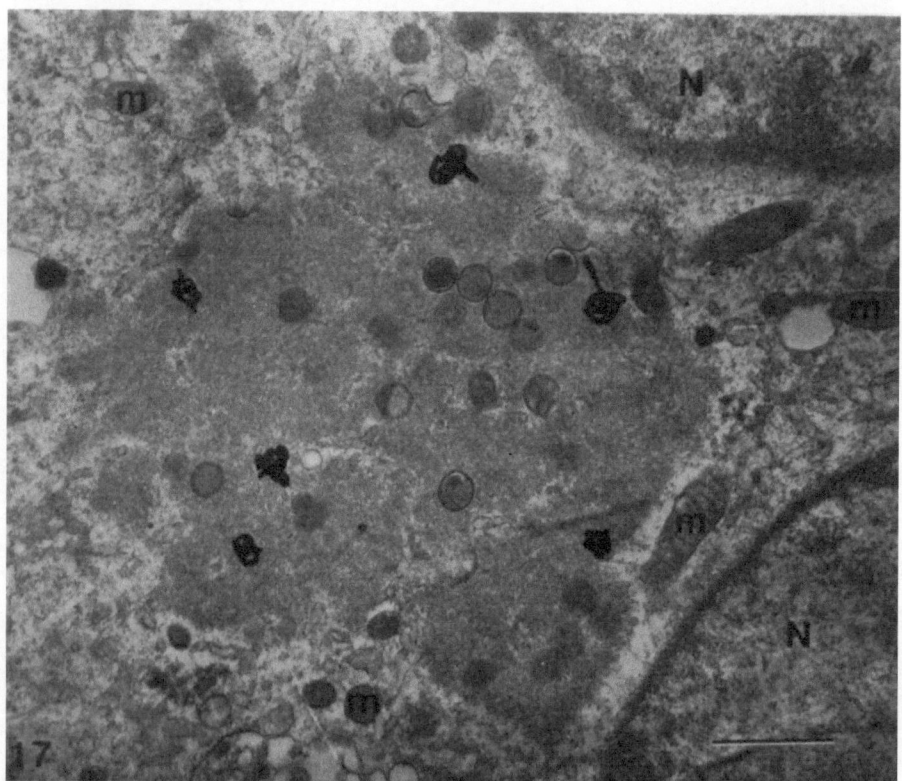

Fig. 17. Electron micrograph of a typical factory in the cytoplasm of a mouse fibroblast sampled 6 hours post inoculation with vaccinia virus. The autoradiogram reveals the presence of labeled DNA from the inoculum within the viroplasmic matrix or "factory" inside of which developmental stages of progeny virus are evident. *N* nucleus, *m* mitochondria × 17,000. (From DALES, 1965b)

Inoculum particles undergoing controlled, sequential dismemberment during penetration have also been characterized morphologically and biochemically after recovery from cell fractions (JOKLIK, 1964; HOLOWCZAK, 1972). These studies support and supplement the electron microscopic observations discussed above. In particular, studies by JOKLIK with vaccinia virions radiolabelled in the DNA and protein moieties have shown that the bulk of the protein remains in the form of large particulates although most of the DNA becomes amenable to DNAse during uncoating, in excellent agreement with the cytological evidence for conversion of cores to shells (DALES, 1965b). Complementary studies, using inhibitors of transcription and translation, which block release of DNA from cores, led JOKLIK (1964) to conclude correctly that ongoing mRNA and protein syntheses are required for poxvirus uncoating. However, the idea that synthesis of an uncoating factor is under the direct control of the host genome has not been substantiated. Instead the critical discovery by KATES and MCAUSLAN (1967a) and MUNYON et al. (1967), of a DNA-dependent RNA polymerase in the virion core provided a basis for explaining how poxviruses may control their own uncoating. The initial and subsequent studies (KATES and BEESON, 1970a, b) clearly demonstrated that poxvirus cores in their coated state are able to transcribe the genomes within them, suggesting that the core is where the mRNA specifying the uncoating protein originates. This idea is further supported by the findings that suppression of host-specified transcription by pretreatment with actinomycin D does not prevent uncoating (MAGEE and MILLER, 1968), but application of the antiviral substance interferon does (MAGEE et al., 1968; LEVINE et al., 1967), presumably because a virus-specified function provided by transcripts from cores is repressed in the latter instance (BIALY and COLBY, 1972).

Treatment of cells for 2 or 3 hours with rapidly acting inhibitors of protein synthesis during virus penetration irreversibly prevents DNA uncoating of vaccinia virus cores (DALES, 1965c; Moss and FILLER, 1970), indicating that the nascent mRNA for the uncoating protein may have a relatively short half-life. One study has provided pertinent, but so far unsubstantiated data purporting to show induction within infected cells of an in vitro "decoating" enzymatic activity with a half life of only a few hours (ABEL, 1963). It is worth mentioning that under some circumstances, as with Molluscum contagiosum infections of primate cells, although there is transcription in inoculum cores, uncoating does not occur for reasons that remain obscure (MCFADDEN et al., 1979).

Our current ideas about poxvirus uncoating conceptualized in terms of molecular biology provide satisfactory explanations for genetic phenomena associated with recombination, marker rescue and the reactivation initially described by BERRY and DEDRICH (1936). With respect to reactivation involving poxviruses of different serotypes, although one partner in the dual infection is thermally inactivated and the genome of the other is rendered incapable of replication by treatment with nitrogen mustard or by limited ultraviolet irradiation (JOKLIK et al., 1960a, b), replication of the genome of the heat-treated virus occurs. One may now suppose that transcription from the partner with an impaired genome provides mRNA for uncoating the heat-denatured companion poxvirus.

V. Transcription and Translation

Unique features in the replication scheme of poxviruses, exemplified by vaccinia virus, which include cytoplasmic DNA replication and involvement of core enzymes, have generated wide interest in these agents as models for investigating phenomena of transcription and translation pertinent to the eukaryotes and their viruses. Such research activities are reflected in numerous publications from which a fairly clear and detailed understanding of RNA and protein syntheses has emerged.

Transcription and translation are generally divisible among DNA agents with respect to DNA replication into prereplicative or *early* and postreplicative, *late* phases. These broad divisions have been subdivided further into a class of *immediate-early* functions related to transcription from the core, *early* functions expressed after uncoating, *late* ones related to materials involved in virion assembly and maturation and *late-late* functions. The last class is associated with virus-specified products, such as the hemagglutinin, which are not components of the virion itself, but are controlled in a manner that brings about their transcription and translation into polypeptides or glycoproteins after virion maturation is already well underway (ICHIHASHI and DALES, 1973).

A. Early Transcription

The significance of data suggesting that transcription may occur in the virion core (MUNYON and KIT, 1966) was clarified by the discovery of a DNA-dependent RNA polymerase in cores of vaccinia virus by KATES and McAUSLAN (1967a), and MUNYON *et al.* (1967), which was corroborated later (KATES, 1970; WOODSON, 1967). Subsequent detailed investigations demonstrated that transcription from the core involves the simultaneous initiation of about 50 to 100 RNA chains and their extrusion through the proteinaceous coat of the core by an ATP-dependent process (KATES and BEESON, 1970a, b); PAOLETTI and LIPINSKAS, 1978b). Since each mRNA chain has a stretch of 50 to 200 polyadenylate [poly(A)] residues covalently linked with it (KATES and BEESON, 1970b), it was initially presumed that poly(A) was the transcript of poly(dT) sequences in the vaccinia genome. However, studies that followed showed that poly(A) is added to the 3' end of the mRNA chain (SHELDON *et al.*, 1972; NEVINS and JOKLIK, 1975) only after transcription has been terminated (SHELDON, and KATES 1974), in a reaction catalyzed by a terminal riboadenylate-transferase or poly(A) polymerase activity (MOSS *et al.*, 1975; MOSS and ROSENBLUM, 1974). Thus, mRNA synthesis and polyadenylation utilize independent activities, both of which have been isolated and purified to varying degrees (BAROUDY and MOSS, 1979; SPENCER *et al.*, 1979; NEVINS and JOKLIK, 1975). Each activity possesses the same physical, chemical and biological characteristics, including its polypeptide constitution, whether it is derived from cores of vaccinia or from cytoplasmic extracts of infected cells (NEVINS and JOKLIK, 1977a, b). Some concern remains, however, as to the native state of these polymerases following isolation, because the enzymes, while still in the core, function optimally with Mg^{++}, while after isolation they act maximally in the presence of Mn^{++}. However, the overall impression remains that the synthesis of both immediate-early mRNA arising from cores and early mRNA produced following uncoating in the cytoplasm is catalysed by the same enzyme originating from the inoculum.

Another feature common to the mRNA molecules of eukaryotes and their viruses is modification at the 5'-end, in the form of methylated *caps*. The enzymatic

activities involved in capping of immediate early transcripts have been identified
in cores of vaccinia and characterized, as described in section II.G. Commencing
with the original discovery by URISHIBA et al. (1975), the sequence of steps in the
capping process and enzymatic activities involved have been elucidated (WEI and
MOSS, 1974; TUTAS and PAOLETTI, 1977; MOSS et al., 1976; BOONE and MOSS, 1977;
NUSS and PAOLETTI, 1977; BOSSART et al., 1978b). Nascent 8 to 12s molecules
emerging from the core bear the capped structures $m^7G(5')pppN_1^m$—N_2^m, in
which A^m and G^m predominante at position N_1 (WEI and MOSS, 1975). In comparing caps on early and late mRNA it was found that the early species contain more
G^m than A^m and $m6A^m$ at the N_1 position and overall are methylated more
extensively at N_2 than the late species. The significance of these findings in terms
of regulatory mechanisms remains to be determined.

Concerning the existence of higher molecular weight mRNA precursors, large
transcripts of 20 to 30s have been identified in vaccinia cores and shown by
appropriate hybridization-competition experiments to be precursors of the 8 to
12s extruded mRNAs (PAOLETTI, 1977a, b). Supporting evidence comes from
experiments utilizing analogues of ATP, in the presence of which preformed
20 to 30s RNA is neither cleaved nor transferred from the core. Similar inhibition
of vaccinia virus mRNA processing is observed during hyperthermia, although
capping and polyadenylation occur normally at 55 to 57° (HARPER et al., 1978).
The precursor molecules are methylated but not polyadenylated and apparently
are devoid of non-informational sequences or 'introns' related to the splicing
process. One might, therefore, conclude that during immediate-early transcription,
polycistronic RNA molecules are produced which become simultaneously cleaved,
polydenylated and extruded from the core as functional, monocistronic mRNAs
(PAOLETTI and LIPINSKAS, 1978). However, UV transcriptional mapping (BOSSART
et al., 1978b) suggests that only a single message is encoded in each 20 to 30s
precursor (see section V.C.).

B. Early vs Late Transcription

As was to be anticipated from the observed rapid shut-off of host protein
synthesis, the polyribosomes formed after vaccinia infection contain virus-
specified mRNA, identified initially by base-composition analysis (BECKER and
JOKLIK, 1964), and originating undoubtedly from the virosomes or cytoplasmic
"factories" (DAHL and KATES, 1970). Inhibitors of transcription, such as actinomycin D, bring about dispersal of such polyribosomes, thereby causing at least a
fraction of the rapidly turning-over nascent mRNA (SHATKIN, 1963) to remain
linked to monoribosomes or to the smaller ribosome subunit (SHATKIN et al., 1965).

Analyses by hybridization-competition or by hybridization kinetics revealed
the occurrence of temporally separated early and late mRNA sequences (ODA and
JOKLIK, 1967; DAHL and KATES, 1970). The switchover from early to late mRNA
is never complete and it is highly variable in extent depending on the host cell
involved. For example, switchover from early to late gene expression appears to
be regulated more precisely in HeLa cells than in mouse L strain fibroblasts (ODA
and JOKLIK, 1967). Perturbations caused by physical agents, such as elevated
temperatures (STEVENIN et al., 1969) or sonic oscillations (BOONE and MOSS, 1978)

may cause slackening of transcriptional controls, whereas agents inhibiting uncoating, such as cycloheximide, or inhibiting DNA replication, such as cytosine arabinoside, prevent synthesis of late mRNA (ODA and JOKLIK, 1967; KAVERIN et al., 1975).

Another mechanism for controlling transcription involves regulation of the quantities of various mRNA classes produced at different periods of the replication cycle. During normal conditions prevailing in vivo, the early mRNA sequences expressed account for about 20% of the information content of the vaccinia virus genome, whereas late mRNA represents almost 50% of the genome (KAVERIN et al., 1975; PAOLETTI and GRADY, 1977; BOONE and MOSS, 1978). The mRNA arising by in vitro transcription from cores includes small quantities of late species complementary to 30% of the genome, implying that, in the immediate-early class of mRNA produced in vitro more sequences are represented than appear during transcription in vivo (PAOLETTI and GRADY, 1977). Therefore, transcriptional regulation in vitro must be less precisely controlled.

The class of early mRNA comprises molecules that are generally smaller than those representative of late mRNA (ODA and JOKLIK, 1967; ATHERTON and DARBY, 1974), but polyadenylation and capping occur in both classes to about the same extent (ATHERTON and DARBY, 1974; NEVINS and JOKLIK, 1975; BOONE and MOSS, 1977). The early class of transcripts produced during in vitro transcription from cores is also of low molecular weight and is capped and polyadenylated. The relative complexity of the sequences represented among these RNAs suggests the presence of a large number of efficient promoters throughout the genome (PAOLETTI et al., 1980).

The variety of different mRNA species accumulating during various phases of transcription is ascertained by analysing the coding complexity reflected in the transcripts. Some sequences may be 11 to 43 times more abundant in late mRNA classes compared to early mRNA. Moreover, within the early mRNA species certain sequences may occur in great molar excess (PAOLETTI and GRADY, 1977). Undoubtedly, mapping of poxvirus genomes by restriction endonucleases and localization of the so-called "R" loops by electron microscopy will continue to provide further insights into temporally related regulation of early or late transcription in different segments of the DNA genome. To date BARBOSA et al. (1979), and WITTEK et al. (1980 b), have demonstrated that inverted, terminal repetitions each $\sim 7 \times 10^6$ in MW, contain information coding for early mRNAs of vaccinia virus.

A somewhat puzzling but reproducible finding concerns formation of small quantities of double-stranded (ds) RNA molecules at least 1000 base pairs long as a product of infection. Following denaturation, this dsRNA was shown to be encoded by sequences equivalent to about $\frac{1}{4}$ of the genome, whereas total late vaccinia mRNA is homologous to almost $\frac{1}{2}$ of the genome (COLBY and DUESBERG, 1969). However, information encoded in the dsRNA represents transcripts of each of the DNA fragments created by restriction with the enzyme Hind III, implying that symmetrical transcription can occur along the entire genome, and is not confined only to the terminally-situated repetitions (BOONE et al., 1979; VARICH et al., 1979).

Although regulation of vaccinia virus transcription and translation appears to occur in enucleated cells or cytoplasts, evidence has been published implicating a function for the nucleus in vaccinia virus-specified mRNA synthesis (BOLDEN et al., 1979; KIT et al., 1964). These data, coupled with the observation that virion assembly in cytoplasts is very inefficient, although expression of early functions and DNA replication appear to progress normally (PENNINGTON and FOLLETT, 1974), imply that a host-related nuclear function is essential for replication of poxviruses. Such a function is most probably related to RNA polymerase II (pol. II) of the host as implied by effects produced with α-amanitin (HRUBY et al., 1979; SILVER et al., 1979). This toadstool toxin, at appropriately low concentrations, specifically inhibits transcription by pol. II in mammalian cells. Contrary to previous findings, recent evidence shows that α-amanitin interferes with the formation of infectious vaccinia virus progeny when host cells contain pol. II sensitive to the drug (SILVER et al., 1979), but interference is not observed in a mutant cell which possesses a pol. II resistant to the toxin. Concerning intranuclear transcription from the host genome, circumstantial evidence indicates that γ-irradiation prior to infection, sufficiently intense to make the host nucleus transcriptionally dysfunctional, fails to reduce vaccinia virus replication (SILVER et al., 1979). This information, coupled with the reported formation of a few mature progeny vaccinia virions in cytoplasts, leads one to hypothesize that pol. II participates either alone or complexed with virus transcriptase to generate late transcripts from the vaccinia virus genome. This is in contrast with expression of early functions, which are catalyzed exclusively by the virion transcriptase introduced with the inoculum.

C. Translation

The capacity of poxviruses, exemplified by vaccinia virus, to inhibit host-specified protein synthesis rapidly and efficiently offers important advantages for studies of virus-related translation and its control. The production of large pools of nascent virus protein within 15 to 60 minutes after penetration can be related to the prior appearance of new polyribosomes which contain viral mRNA (BECKER and JOKLIK, 1964), and which are actively engaged in translation, as demonstrated by immunoprecipitation employing suitable antiviral antisera. The products of translation accumulate within factories of CAIRNS (1960) which are also termed viroplasmic foci.

To initiate translation, the mRNA engages the smaller ribosomal subunit, in accord with the classical sequence applicable to the eukaryotes (METZ et al., 1975a; JOKLIK and BECKER, 1965; SCHARFF et al., 1963). Virus-specified protein factors involved in assembly and functioning of poxvirus-related polyribosome complexes remain to be discovered.

The significance of the modifications present at the 3'- and 5'-ends of the mRNA has been the subject of numerous investigations employing cell-free translation, usually with extracts of either wheat germ or reticulocytes. In such systems, the role of poly (As) at the 3' end remains enigmatic. Their presence on the mRNAs, whether as full length or short chains or, indeed, their absence altogether does not appear to affect the rate of translation (NEVINS and JOKLIK, 1975). By contrast, methylation and capping at the 5' end were shown to be obligatory for the efficient translation of vaccinia virus mRNA (WEBER *et al.*, 1977; BOSSART *et al.*, 1978a). The critical role of the caps in binding of mRNA to the 40s ribosomal subunit was demonstrated by the ability of a capped high MW RNA precursor derived from virion cores to bind to this subunit, although less efficiently than the 8 to 12s extruded mRNA (BOSSART *et al.*, 1978a).

Further information about the specificity and control of virus translation *in vitro* comes from numerous studies (JAUREGUIBERRY *et al.*, 1975), some involving cell-free systems in which transcription is coupled with translation in a single reaction mixture (PELHAM, 1977). If isolated virus cores are placed into wheat germ or reticulocyte extracts containing the appropriate metabolites, nascent mRNA is synthesized from the cores and thereupon immediately translated (COOPER and MOSS, 1978, 1979). In one study, the coupled system synthesized over 20 polypeptides authenticated as belonging to vaccinia virus-specified early functions by comparisons of tryptic peptide band patterns with the patterns of selected proteins made *in vivo*.

To ascertain whether the frequency of capping can be equated with the number of promoters occurring on vaccinia mRNA, one study employed the so-called "UV transcription mapping" procedure (BOSSART *et al.*, 1978b) whereby the target size of the DNA template for an mRNA is estimated from the UV dose required to eliminate the protein product of the mRNA in a translation system. Data from this investigation suggest that only a single promoter is involved in the formation of the high MW 20 to 30s precursors in the vaccinia core, although this RNA does not appear to function as a polycistronic message. Therefore, molecules of the extruded 8 to 12s mRNAs, which function as monocistronic messages, appear to be formed by cleavage of the 20 to 30s precursor and capping as well as polyadenylation of each fragment arising therefrom (COOPER and MOSS, 1978).

It is reassuring to find that the appearance of defined classes of poxvirus mRNAs in a temporal sequence can be matched with the synthesis of defined groups of proteins (ESTEBAN and METZ, 1973; BODO *et al.*, 1972; BAGLIONI *et al.*, 1978). Such nascent proteins are readily identifiable by immunoprecipitation (MOSS and SALZMAN, 1968; MOSS and KATZ, 1969; KATZ and MOSS, 1969; SALZMAN and SEBRING, 1967; COHEN and WILCOX, 1966, 1968), immunofluorescence (LOH and RIGGS, 1961), or polyacrylamide gel electrophoresis (PAGE) coupled with autoradiography. Rapid cessation of host translation, irrespective of viral DNA replication (MOSS and SALZMAN, 1968; ESTEBAN and METZ, 1973), enhances the sensitivity of these analyses. Polypeptides characterized also as antigens of both the early and late varieties become concentrated in the factories (ESTEBAN and METZ, 1973; KATZ *et al.*, 1974). Analysis by one-dimensional PAGE could detect about 30 early and 50 late proteins in such factories (PENNINGTON, 1974). To ascertain the positions on the vaccinia virus map of the individual functions, isolated EcoRI or HindIII restriction fragments of the genome were cloned for the purpose of hybrid selection of individual mRNAs which the fragments encode, i.e. transcriptional mapping. The selected mRNAs forming hybrids with about 50% of the genome were employed in an *in vitro* translation system and shown to encode over 70 early and more than 40 late polypeptides (MORGAN and ROBERTS, 1980; MAHR and ROBERTS, 1980; BELLE ISLE *et al.*, 1980; all in appended bibliography).

The usual switch from synthesis of early to late polypeptides associated with the normal cycle of vaccinia virus production may be interrupted by inhibitors of DNA replication such as cytosine arabinoside or hydroxyurea, resulting in an indefinite prolongation of the synthesis of many early proteins (PENNINGTON, 1974). Similarly, when virus DNA is synthesized in the presence of analogues and contains substituted halogenated bases such as bromodeoxyuridine, normal expression of the polypeptide spectrum is disrupted without affecting specifically the synthesis of all late proteins (PENNINGTON, 1976).

In the category of virus-coded non-virion materials are products of both early and late functions. Among the former is an antigen(s) associated with the plasma membrane (UEDA et al., 1969; BALL and MEDZON, 1976), perhaps relevant to virus-induced alteration in permeability properties of that membrane (BALL and MEDZON, 1973). The late-late functions comprise another membrane-associated antigen, the hemagglutinating glycoprotein (WEINTRAUB and DALES, 1974), as well as a cytoplasmic polypeptide that accumulates into the large aggregates, termed A-type inclusions (ATI), of cowpox infected cells (ICHIHASHI and DALES, 1973). The ATI are formed by a unique translation scheme in which polyribosomes encrusting the surface of the inclusion probably synthesize the protein out of which a continuously increasing mass of ATI is formed, as illustrated in Fig. 23A.

Interferon, one of the naturally occurring cellular factors induced by poxvirus infection, can interrupt translation. In cells treated with interferon, vaccinia virus-specified translation is blocked. Coincidentally, the polyribosomes become dispersed (METZ et al., 1975; JOKLIK and MERRIGAN, 1966; KERR et al., 1974; BODO et al., 1972) and initiation steps involving binding of mRNA to the 40s ribosomal subunit fail to occur, both in vivo (METZ et al., 1975), and in a cell-free system (KERR et al., 1974). Accumulation of dsRNA synthesized by symmetrical transcription of the viral genome might provide the induction mechanism for interferon production (COLBY and DUESBERG, 1969). Since, however, interferon exerts its influence at the external cell surface, the factor(s) disrupting translation of viral proteins in the original infected cell cannot be interferon itself. One presumption is that the dsRNA influences translation directly, by interfering with binding of formyl-methionine-transfer RNA (tRNA) to the mRNA, as demonstrable in the wheat germ cell-free translation system (BAGLIONI et al., 1978; METZ et al., 1975). Coincidentally, a virus-specified protein kinase may bring about phosphorylation of serine and threonine residues in polypeptides S2 and S16 of the 40s ribosomal subunit. This protein kinase could be brought into the host along with the inoculum as a component of the core. This notion stems from the observation that virus-induced ribosome protein phosphorylation can occur following UV irradiation of the virus inoculum or even after uncoating of the penetrating inoculum is blocked with cycloheximide (KAERLEIN and HORAK, 1976, 1978).

By manipulating the critical concentration of essential basic amino acids, such as arginine, ARCHARD and WILLIAMSON (1971) elucidated the control and expression of the synthesis of early or late virus-specific basic polypeptides and polyamines. One major component, which accounts for about 10% of the virion mass and which requires arginine, is a histone-like protein of 10—11K MW, containing phosphorylated serine and threonine residues (POGO et al., 1975;

unpublished data). Synthesis of this polypeptide is coordinated with DNA replication, placing it in a class intermediate between early and late proteins.

Synthesis of the polyamines, among them the virus-specified spermidine (WILLIAMSON, 1976), which associate intimately with the DNA, neutralizing its acidic charge, also depends upon the availability of relatively high arginine concentrations. Induction of a vaccinia virus-specified enzyme, arginino-synthetase lyase, catalysing the conversion of citrulline into arginine, may be an essential

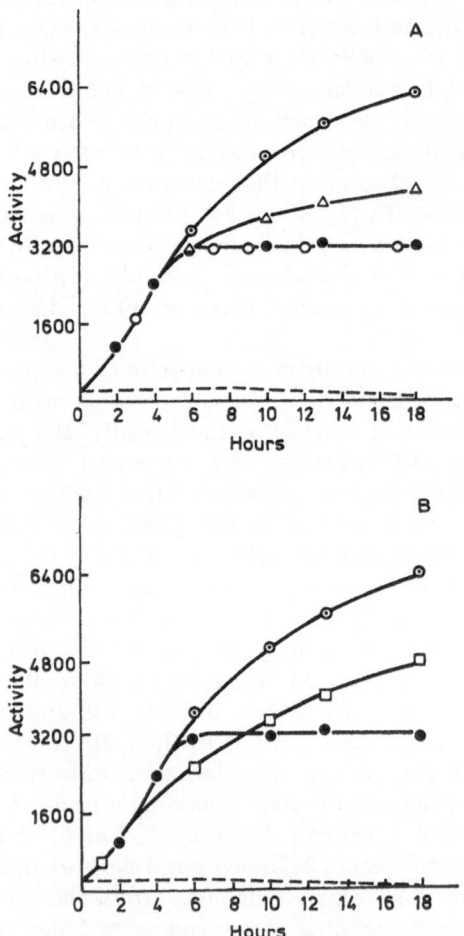

Fig. 18. The effect of actinomycin D on TdR kinase synthesis of CP infected cells. A the effect of different concentrations of actinomycin added at 3 hours after virus infection. •————• No actinomycin added. ○————○ 0.25 µg actinomycin per ml; △————△ 1.0 µg actinomycin per ml; ⊙————⊙ 5.0 µg actinomycin per ml; — — — uninfected cells plus 5 µg actinomycin per ml. B The effect of adding actinomycin (5 µg/ml) at various times after infection. •————• No actinomycin added; — — — actinomycin added at the time of infection; □————□ actinomycin added at 1 hour after infection; ⊙————⊙ actinomycin added at 2, 3, or 4 hours after infection. (From McAUSLAN, 1963)

step in spermidine synthesis (WILLIAMSON and COOKE, 1973). Formation of putrescine, an intermediate in polyamine synthesis, requires induction of another enzyme, ornithine decarboxylase, which participates in the ornithine-citrulline synthetic cycle (HODGSON and WILLIAMSON, 1975). After early induction, there is a late repression of this enzyme following DNA replication. However, the late repression control is abolished in the absence of DNA synthesis, causing an over-production of the enzyme. A similar type of regulation pertains to another early enzyme, thymidine kinase (KIT et al., 1963d, 1964), which presumably represents one of the immediate-early functions expressed by transcription from the core. This enzyme participates in the scavenger pathway of DNA synthesis, converting deoxyribonucleosides into the respective triphosphates (McAUSLAN and JOKLIK, 1962; McAUSLAN, 1963a). Induction of thymidine kinase after vaccinia virus infection is evident within 2 hours and repression occurs by 6 hours, unless late mRNA transcription is suppressed by inhibitors such as actinomycin D (Fig. 18). A similar regulation prevails when this enzyme is induced by Shope fibroma virus infection (BARBANTI-BRODANO et al., 1968). The mRNA for thymidine kinase is stable for at least 18 hours under conditions blocking repression (McAUSLAN, 1963b). Virus specification of the thymidine kinase is shown by the characteristics of the MW of the polypeptide (KIT et al., 1977), antigenic specificity, lower themosensitivity and lower Km, all of which are different from those of similar host enzymes.

Poxvirus-specified enzymes that function in the synthesis, modification or hydrolysis of DNA are represented among both early and late functions (ERON and McAUSLAN, 1966). The DNA polymerase and polynucleotide ligase activities might be placed in a temporally intermediate class of functions because their induction occurs at the time of DNA replication (SAMBROOK and SHATKIN, 1969; MAGEE and MILLER, 1967; KATES and McAUSLAN, 1967b). An alkaline exonuclease with specificity for dsDNA appears as an early function (ERON and McAUSLAN, 1966), while the core-associated endonuclease with a neutral pH optimum, and an acidic exonuclease, both with specificity for ssDNA templates, belong in the late category (POGO and DALES, 1969b; McAUSLAN and KATES, 1967; SAHU and MINOCHA, 1974). The virion-associated RNA polymerase, although involved in both immediate-early and early transcription, is induced late, at the time of virus assembly and maturation (KATES et al., 1968; PITKANEN et al., 1968; NAGAYAMA et al., 1970). Another activity present in the core, the nucleotide phosphohydrolase, is also expressed as a late function (POGO and DALES, 1969b). However, some other core enzymes involved in the methylation and capping of mRNA, including polynucleotide 5' phosphatase (TUTAS and PAOLETTI, 1978), mRNA guanylyl transferase and mRNA methyl transferase(s) (BOONE et al., 1977), are apparently induced as early functions. Thus, enzymes of the virus core which act soon after penetration, during the early stages of the replication cycle, are synthesized either early or late after infection.

The existence of tight transcriptional control of the poxvirus genome is evidently fundamental to the orderly appearance of structural and functional polypeptides required for the replicative process.

VI. Assembly and Morphogenesis

The temporal sequence of events in poxvirus morphogenesis can be correlated with the appearance of early and late classes of virus polypeptides as characterized by their PAGE gel profiles (Moss and SALZMAN, 1968; PENNINGTON, 1974) and their antigenic properties (LOH and RIGGS, 1961; SALZMAN and SEBRING, 1967; WILCOX and COHEN, 1967; COHEN and WILCOX, 1968). The virions are assembled in specific cytoplasmic foci of viroplasm out of pools of viral DNA, the requisite virus-specified proteins and host-derived lipids (CAIRNS, 1960; LOH and RIGGS, 1961). The reconstructed sequence of assembly stages is based upon numerous complementary biochemical, genetic and structural studies (Figs. 19 A—I), carried out mainly with the prototype vaccinia virus. However, the scheme most probably represents a basic pattern pertinent to all viruses in this Family.

A. The Envelope

Poxvirus envelopes are external lipoprotein teguments organized into the classical 50 to 55 nm bilayer-membrane form, which initially surround spherical immature particles and later enclose mature progeny virions (DALES, 1963; DALES and MOSBACH, 1968). During a synchronized one-step growth cycle the envelopes are the first identifiable virion structures assembled within factories, in line with suggestive data from the initial electron microscopic studies obtained with vaccinia virus (DALES, 1963; GAYLORD and MELNICK, 1954; MORGAN et al., 1954; VALLEJO-FREIRE et al., 1957/58), and subsequently with other orthopoxviruses (BERGOIN et al., 1969; DALES and BERGOIN, 1971; GRANADOS, 1973; GRANADOS and ROBERTS, 1970; STOLZ and SUMMERS, 1972; TSUHURA, 1971; TSURUHARA and

---

Fig. 19. Selected examples of vaccinia virus morphogenesis in relation to the self-assembly and modifications of the envelope. *A* Early in development where the unit membrane is coated by a well defined external layer of spicules (arrows). The segment of curved envelope appears rigid, but in the absence of spicules it is flexible. × 225,000. (From SCHWARTZ and DALES, unpublished.) *B* A complete, spherical envelope encloses the nucleoprotein fibrous matrix of an immature virion. × 80,000. *C* and *D* Immature particles contain dense, DNA nucleoids. × 90,000. (From KAJIOKA et al., 1964.) *E* Transitional stage of development when internal reorganization is evident. × 90,000. *F* Maturing or mature virion contains two lateral bodies and a core enclosed by a thick membrane. The envelope is devoid of spicules. Such particles are usually evident 4 to 5 hours post inoculation. × 90,000. (From DALES and MOSBACH, 1968.) *G* An unusual configuration of envelopes in early development. The presence of spicules on both faces of adjacent membrane bilayers may have caused a constraint of the normal curvature of the envelope. × 120,000. (DALES, unpublished.) *H* Developmental defects in the case of mutant *ts* 6757 result in self-assembly of very long segments of spicule-coated envelope (arrows), which frequently become folded into multilayers. Such envelopes are spatially independent of the DNA-protein matrix synthesized by this mutant. × 123,000. (From DALES et al., 1978.) *I* Resculpturing of the envelope during maturation. When differentiation into a core and lateral bodies is evident, the spicules (long arrows) are removed or lost from the external envelope (short arrows), usually on the surface opposite from the "factory". Such transitional forms in vaccinia virus development are encountered rarely with wild type virus, but accumulate in large number in the case of a slowly maturing mutant like *ts* 1911. × 142,000. (From DALES et al., 1978)

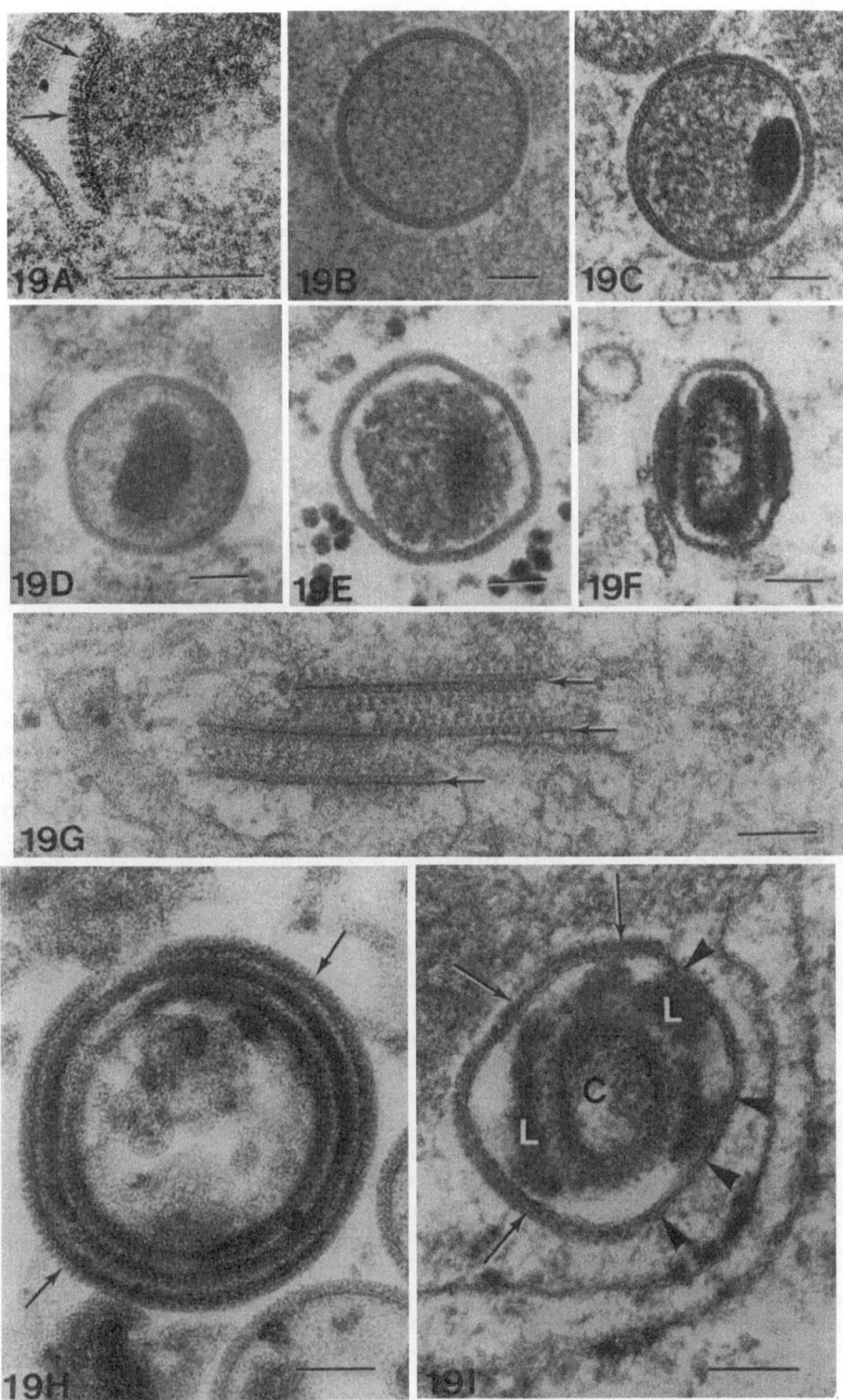

Fig. 19 A—I

TSURUHARA, 1973). When rapidly acting metabolic inhibitors are employed, it can be shown that the formation of envelopes around immature vaccinia virions is dependent on transcription at 2 to 2.5 hours and translation at 3 to 3.5 hours post inoculation. These envelopes are assembled sequentially into uniform, spherical particles and the rigidity of their form depends upon an external backing of spicules (DALES and MOSBACH, 1968) (Figs. 19 A—E). When these spicules are missing from the bilayer, particularly as evident after treatment during vaccinia virus infection with the antibiotic rifampicin or in the case of some conditional-lethal assembly mutants of vaccinia virus, the envelope retains its flexibility and pleomorphism (Fig. 19 A) (DALES and MOSBACH, 1968; NAGAYAMA et al., 1970; GRIMLEY et al., 1970). Therefore, the curvature required to produce a spherical envelope must be controlled by a precise attachment of spicules. It is noteworthy that, when two spicule-backed envelope segments become closely apposed by chance, the tendency to adopt curvature is counteracted (Fig. 19 G). Abnormally developing envelopes sometimes fail to become sealed into spheres enclosing material of the viroplasm and are instead assembled into long sheets tightly wound into a multilayered form (Fig. 19 H), as evident with other conditional-lethal, *ts* mutants of vaccinia virus (DALES et al., 1978).

The spicule-backed envelopes are produced as early functions, and can, therefore, be assembled to enclose immature particles lacking viral DNA, when DNA replication is prevented by drugs or nucleoside analogues such as hydroxyurea, fluorodeoxyuridine and cytosine arabinoside (KAJIOKA et al., 1964; ROSENKRANZ et al., 1966; POGO and DALES, 1971). However, if envelopes are formed in the absence of coordinate DNA and late protein synthesis, they most probably cannot subsequently participate in the assembly of mature progeny virions after inhibition of DNA synthesis is reversed. On the other hand, malformed envelopes, assembled around quanta of viroplasmic matrix synthesized during treatment with rifampicin, appear to undergo "repair" after the drug is washed out and they are subsequently utilized in the formation of mature progeny virions (STERN and DALES, 1967 a; NAGAYAMA et al., 1970). These observations indicate that, during normal virus development, envelopes are assembled to enclose the DNA plus early and late proteins constituting the mature virion.

The structure of the external surface of the envelope undergoes specific modification which is temporally connected with conversion of the immature virion into a mature particle, characteristically associated with internal differentiation of the particle and its migration out of the viroplasmic matrix. These modifications involve replacement of the spicule layer by surface tubular elements (STE) and internal assembly of the core with its lateral bodies (STERN and DALES, 1976 b). The first step in late morphogenesis appears to include dissociation of spicules from the bilayer, restoring flexibility to the envelope and thus allowing the change in shape from the spherical immature particle to the brick-shaped mature virus. Transitional stages of envelope modification are particularly clear in the case of an insect poxvirus (STOLTZ and SUMMERS, 1972) and with vaccinia virus *ts* mutants blocked at a step in maturation (STERN et al., 1977) (Fig. 19 I). The STE on mature virions have been isolated in a pure state (Fig. 8), and shown to be comprised of a single 58 K polypeptide species by one-dimensional PAGE. This protein elicits avid neutralizing antibody in rabbits (STERN and DALES, 1976 b). Pulse-chase experiments using isotopically labelled amino acids reveal that during vaccinia virus assembly, STE protein is among the last components to be incorporated into virions (STERN and DALES, 1976a), a finding consistent with the idea that the spicules are displaced during maturation (SAROV and JOKLIK, 1973).

The lipids which are completely extractable from the surface of mature pox-viruses by detergent undoubtedly also exist as a component of the bilayer structure of immature virus envelopes. Experiments utilizing suitable radioactive precursors, such as glycerol, acetate, ^{32}P-PO$_4$ and fatty acids, show that developing particles of vaccinia virus acquire phospholipids of the envelope indiscriminately from both preexisting and nascent molecules synthesized after infection (STERN and DALES, 1974; DALES and MOSBACH, 1968). These data, which include all the species of phospholipid, demonstrate that their synthesis is entirely under host control. However, the proportions of phospholipid species in virion envelopes is somewhat different from their proportions in host cell membranes. For example, the relative concentration of phosphatidylethanolamine is significantly reduced in virus envelopes, and a substance tentatively identified as acylphosphatidylglycerol is present there in several times greater abundance (STERN and DALES, 1974). These findings suggest that virus envelope proteins might control the phospholipid composition during envelope formation. Poxvirus envelopes are assembled within the cytoplasm *de novo*, in contrast to the envelopes of "budding" viruses which are continuous with host cell membranes during development. There must, therefore, exist a mechanism for transferring phospholipids from cellular membranes to nascent envelopes developing inside the factories. Whether such transfer depends upon specific catalytic carriers, termed phospholipid exchange proteins, as sugge-sted by preliminary evidence (STERN and DALES, 1974), or on some other mecha-nism, remains to be established.

Glycolipids, when quantified relative to amounts of prevailing phospholipid, exist in the vaccinia virus envelope in the same molar ratios as in host membranes (ANDERSON and DALES, 1978), implying that glycolipids of cellular origin are incorporated into the unique vaccinia envelope without the type of discrimination evident with phospholipids. However, vaccinia virus infection does cause profound alterations in host glycolipid composition, namely, large increases of the least complex mono- and dihexosylceramides and commensurate reduction in the more complex glycolipids and gangliosides (ANDERSON and DALES, 1978). From this evidence, it can be deduced that rapid changes in glycolipid composition of the host are immediately reflected in the virus progeny, indicating that nascent glycolipids become incorporated into the envelopes of vaccinia virions.

B. Differentiation into Mature Virions

While conversion of spherical immature forms into brick-shaped, infectious virions requires the resculpting of the external surface of the envelope, it also involves the interior conversion of an undifferentiated viroplasmic matrix material into the characteristic core and lateral bodies (Figs. 5, 19F, I). The pattern of development evident with vaccinia virus is quite similar for all orthopox viruses examined to date (BERGOIN and DALES, 1971; DALES, 1963; STOLTZ and SUMMERS, 1972); TSURUHARA and TSURUHARA, 1973). Intermediate stages in the differentiation process have been identified by electron microscopy (DALES and MOSBACH, 1968), and are especially clear when defects in development become evident among *ts* mutants of vaccinia virus (DALES *et al.*, 1978; LAKE *et al.*, 1979). Both DNA replication and expression of late functions are mandatory for the occurrence of internal differentiation, as shown with appropriate metabolic inhibitors (KAJIOKA *et al.*, 1964; ROSENKRANTZ *et al.*, 1966; POGO and DALES, 1971; DALES and MOSBACH, 1968). It is, therefore, noteworthy that blocking transcription at 3 to 3.5 hours and translation at 3.5 to 4 hours, i.e., a time after the bulk of vaccinia virus DNA has been synthesized, does not prevent assembly of immature particles but does inhibit differentiation into mature virions (DALES and MOSBACH, 1968). Two basic models may be considered to explain the interrelationship between formation of the vaccinia virus envelope and the maturation process. According to the first model, the lipoprotein envelopes surrounding immature particles are assembled first and the late polypeptides required for maturation are inserted through them. The second model predicts that only after all requisite components have been incorporated prior to completion and final sealing of the envelope can the assembly and differentiation into mature virions proceed. We shall now consider the evidence favouring the latter alternative.

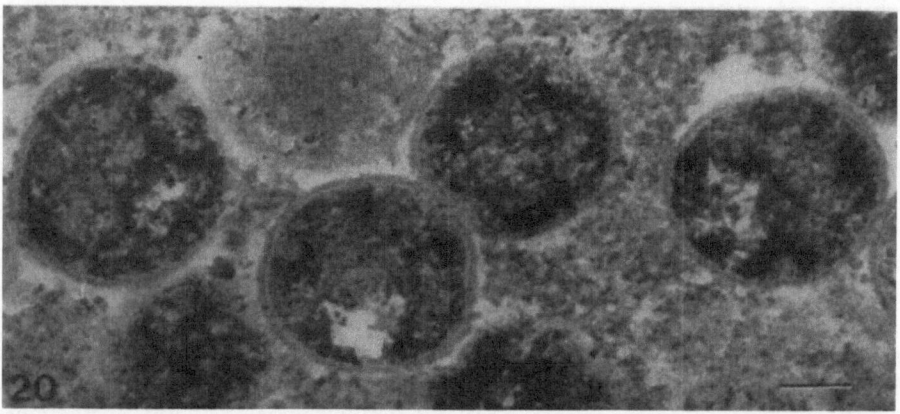

Fig. 20. Section of cytoplasm from a vaccinia virus-infected cell examined histochemically for nucleotide phosphohydrolase. The reaction was carried out as described in Fig. 10. Note the dense lead phosphate product within immature forms of progeny virus and its absence from the envelope. × 100,000. (From GOLD and DALES, 1968)

Since the major envelope polypeptides are among the last to be incorporated during vaccinia virus assembly, envelopment most probably occurs late in the sequence of development (STERN and DALES, 1976a). This view is consistent with evidence showing that envelopes of immature forms assembled during treatment with hydroxyurea to arrest synthesis of DNA and late polypeptides do not participate in the formation of mature progeny after reversal of the block. Following removal of hydroxyurea, DNA synthesis occurs rapidly and is followed by highly synchronous formation of new immature and mature progeny (POGO and DALES, 1971; MORGAN, 1976a, b). Appearance of mature progeny under these circumstances is temporally related to the synthesis of the enzymatic activities of the core (GOLD and DALES, 1968; POGO and DALES, 1969b; POGO and O'SHEA, 1977), including the RNA polymerase, nucleotide phosphohydrolase (NTPase) and two DNAses shown to be late functions (POGO and DALES, 1971). It can be demonstrated cytochemically (Fig. 20) that one of these enzymes, the NTPase, exists within the immature form of vaccinia virus even before the internal reorganization into a core and lateral bodies has been completed (GOLD and DALES, 1968).

It is now evident that, in addition to replication of DNA and appearance of late proteins, specific posttranslational cleavages of these proteins, termed "processing", must occur for completion of virion maturation. During maturation, several precursor polypeptides are simultaneously processed (STERN et al., 1977). They are two major core components, one of which is the 94K precursor which is cleaved to a 62K product and the other a 65K precursor processed into a 60K product (MOSS and ROSENBLUM, 1973). The role of protein processing in poxvirus morphogenesis was initially demonstrated by treatment with β-isatin thiosemicarbazone (EASTERBROOK, 1962), and rifampicin (PENNINGTON, 1973; KATZ and MOSS, 1970a, b). The latter antibiotic also interrupts (a) attachment of spicules to the bilayer of the envelope (NAGAYAMA et al., 1970; PENNINGTON et al., 1970; GRIMLEY et al., 1970), (b) maturation (NAGAYAMA et al., 1970; GRIMLEY et al., 1970) and (c) induction of core-associated late enzymatic activities, including the RNA polymerase, 2 DNAses and NTPase (NAGAYAMA et al., 1970), but it does not affect synthesis of DNA or most of the early and late proteins. Evidence that the effects on development observed are at least partly independent of the transcription-inhibiting influence of rifampicin stems from electron microscopic studies of virus assembly (SUBAK-SHARPE et al., 1969; NAGAYAMA et al., 1970). Interruption of the developmental process is also brought about by elevated temperature, as shown in the case of variola virus infecting chick embryo fibroblasts (COOPER and BEDSON, 1973). At high temperature, viral DNA and some late proteins are made, but induction of the virion RNA polymerase and maturation are blocked. These observations reveal that structural and functional changes associated with poxvirus development and maturation involve closely interrelated events occurring within enveloped virions. This notion is further supported by experiments using a group of vaccinia virus ts mutants (DALES et al., 1978; LAKE et al., 1979; STERN et al., 1977), which mimic almost exactly the above mentioned (a)—(c) effects produced by rifampicin treatment. Upon shift-down of the temperature from 40° to 33°, normal development is restored in the manner evident after washing out rifampicin.

. After reversal of rifampicin inhibition or upon shift-down to the permissive temperatures in the case of ts mutants, additional transcription and translation are required for processing of viral proteins, induction of core enzymes and occurrence of morphological differentiation. However, application of RNA and protein synthesis inhibitors prior to drug or temperature reversal does not affect attachment of spicules and change to the spherical conformation of envelopes (NAGAYAMA et al., 1970; STERN et al., 1977). This finding implies that a pool of spicules exists under non-permissive conditions and envelope self-assembly does not require the prior induction of a nascent protease or factor connected with proteolysis. Since, however, an association between the envelope and spicules, followed by dissociation of spicules from the bilayer, must precede all phenomena connected with differentiation and maturation, it can be assumed that events connected with envelope self-assembly are paramount during the morphogenetic cycle of poxvirus development. The protease(s) active in vaccinia virus formation has a chymotryptic specificity, judging by the fact that protease inhibitors like TPCK and ZPCK block processing of vaccinia preproteins and virus maturation (SILVER and DALES, unpublished). It is uncertain at the time of writing whether the alkaline protease identified in virion cores (ARZOGLOU et al., 1979), is connected with processing phenomena and development.

In summary, the best evidence from the many correlative biochemical and cytological investigations cited above strongly indicates that envelopment, post-translational processing, induction of several late core enzymes, internal re-organization of structure, resculpting of the external surface, and maturation into infectious progeny are interrelated processes in vaccinia virus self-assembly which must occur in a tightly-coupled functional and temporal order.

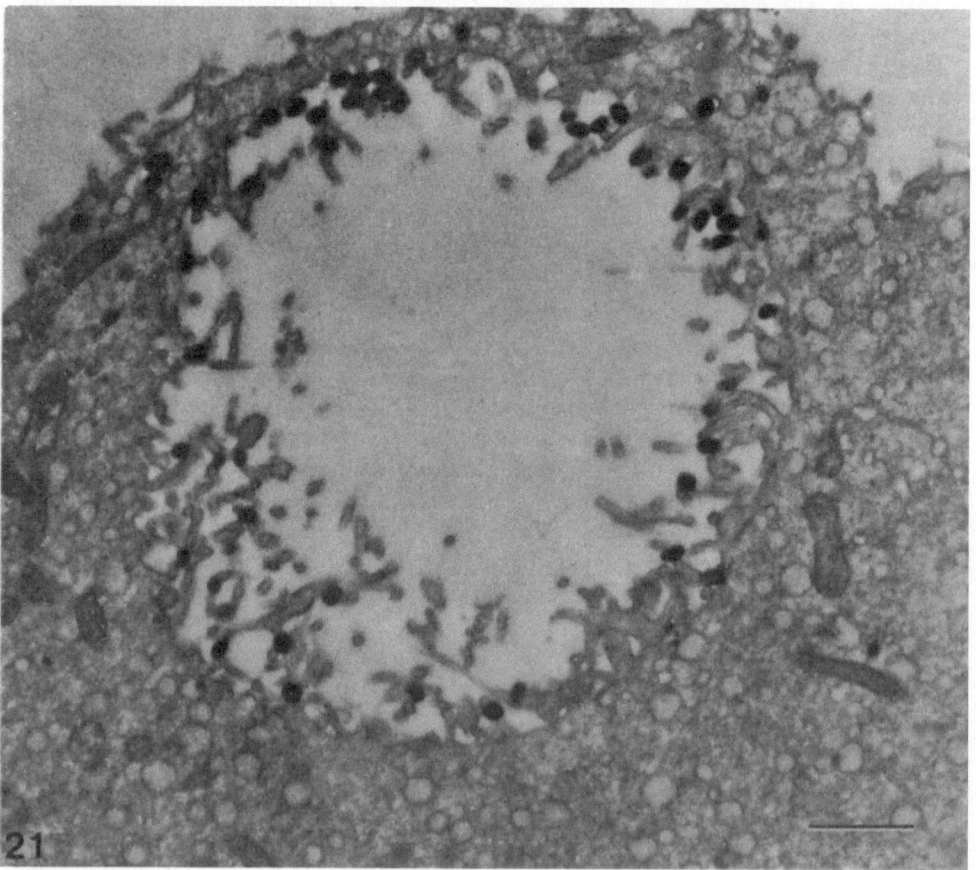

Fig. 21. Section through an invagination near the cell surface of a mouse L strain fibroblast sampled 24 hours after inoculation with vaccinia virus. Numerous microvilli projecting into the luminal surface bear virus progeny. × 12,000. (From DALES and SIMINOVITCH, 1961)

VII. Dissemination from Host Cells

Poxviruses, like certain other animal viruses, have evolved various alternative schemes for efficiently disseminating their progeny which utilize the membrane network of the host, or which depend upon the formation of specific containers in the form of proteinaceous cytoplasmic inclusions.

A. Involvement of Cell Membranes

When cellular membranes are utilized, the mature virions undergo orderly transfer towards and then through the plasma membrane. In the least complex and most direct process, virions migrate through the cytoplasmic matrix towards the cell periphery. After reaching the cell surface, individual progeny particles may become positioned at the tips of long microvilli-like extensions, which are evident by both light and electron microscopy (Fig. 21) (ROBINOW, 1950; DALES and SIMINOVITCH, 1961; STOKES, 1976). Most probably, release to the extracellular *milieu* occurs by breakdown of the extensions at their tips. Immunocytochemical studies indicate that cytoskeletal elements may be involved in egress through microvilli (HILLER *et al.*, 1979). In the case of certain insect orthopoxviruses, release by an analogous type of budding may occur, although by short evaginations of the plasma membrane, instead of through defined microvilli (GRANADOS, 1973; STOLZ and SUMMERS, 1972).

Vectorial transfer of mature progeny of vertebrate poxviruses may also occur by a more elaborate mechanism, whereby membranes of the Golgi complex are utilized for intracytoplasmic wrapping of individual virions (MORGAN, 1976a; ICHIHASHI *et al.*, 1971). The illustrated sequence in Fig. 22A—D of wrapping followed by release at the surface reveals the intricacy of the process. Initially, several Golgi vesicles make contact *via* the cytoplasmic faces of their membranes with the envelope surfaces of mature virions. This interaction presumably involves specific molecular signals of recognition. Fusion or coalescence between the Golgi vesicles ensues to form a continuous sac or cisterna investing individual virions. This fusion process may be controlled by a component of the virion. The wrapped virus particles then migrate to the cell surface where apposition between the outer cisternal membrane and the plasma membrane elicits membrane to membrane fusion at the point of contact, exposing the inner cisternal membrane to the cell exterior. In some instances, membranes on the cytoplasmic side of the cisterna possess dense regions, like those associated with coated vesicles containing the substance clathrin. Such vesicles are believed to participate in the shuttling of materials between the Golgi complex and the cell surface. Thus, when fusion between the smooth and coated vesicles occurs to form a continuous wrapping membrane, a marker becomes available to follow the origin and disposition of the cisternal membranes. During the last stage of egress, the virion is externalized while still enclosed within the residual inner cisternal membrane. This membrane often remains intact after release, but sometimes it becomes ruptured (ICHIHASHI *et al.*, 1971; DALES, 1971; MORGAN, 1976; PAYNE and NORRBY, 1976; PAYNE, 1978). Analyses by PAGE and immunological methods reveal that wrapping membranes derived from the Golgi are modified following infection with vaccinia

virus by addition to them of eight or more virus-specified proteins and glycopeptides, none of which is like any species found in the virion (APPLEYARD *et al.*,
1971; PAYNE and NORRBY, 1976; PAYNE, 1978, 1979). Among vertebrate poxviruses, efficiency of egress by means of Golgi cisternae may be controlled primarily
by the type of host cell (PAYNE, 1979), but the virus strain involved is also a
determining factor. Dissemination by wrapping has not been documented among
insect poxviruses. After wrapping has ceased during late phases of the replication
cycle, or in circumstances not conducive to budding *via* microvilli, the progeny
accumulate in large numbers throughout the cytoplasm as naked virions devoid
of a host membrane integument, frequently becoming lodged at the periphery
subjacent to the cell surface. Such particles become attached to the plasma
membrane itself, as evident after isolation of the virus-membrane complex
(Fig. 23) (WEINTRAUB and DALES, 1974). The trapped naked particles usually
constitute the major fraction of progeny but, to be unmasked as infectious units,
they must be liberated by mechanical disruption of the host cells (APPLEYARD
et al., 1971; BOULTER and APPLEYARD, 1973; EASTERBROOK, 1961).

Virus-controlled modifications of cell membranes, exemplified by induction of
the hemagglutinin (HA) glycopeptide synthesized as a late-late function and

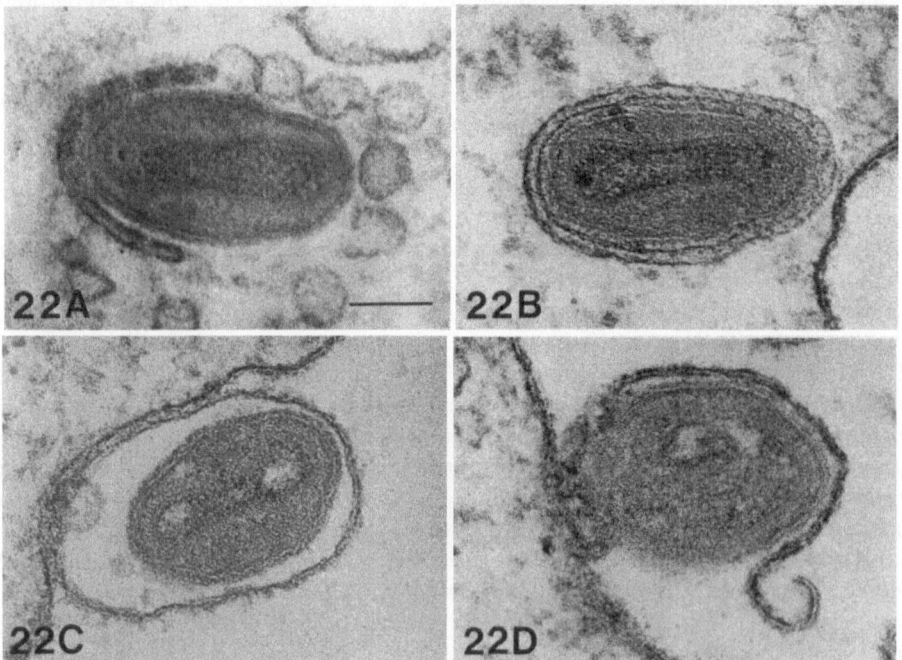

Fig. 22. Selected examples show wrapping of mature vaccinia-virus progeny within
Golgi vesicle membranes. *A* All of the vesicles have not fused into a continuous cisterna. *B* A virion, completely enclosed in a double-membrane cisterna, has migrated
to the vicinity of the plasma membrane. *C* An extracellular particle is enclosed within
the remaining inner membrane of the cisterna. The outer membrane coalesces or fuses
with the plasma membrane during egress. *D* The inner wrapping membrane on the
extracellular virion has become ruptured. × 120,000. (From DALES, 1971)

attached externally to the plasma membrane (BLACKMAN and BUBEL, 1972; WEINTRAUB and DALES, 1974; ICHIHASHI and DALES, 1971; ICHIHASHI et al., 1973), may also influence the cell-to-cell spread of progeny. However, spontaneous variants of vaccinia and variola viruses arising in animals and in culture, which are designated HA⁻ because of their inability to induce HA, have been identified repeatedly (ICHIHASHI et al., 1971; KAKU and KAMAHORA, 1964; CASSEL and FATER, 1958; TSUCHIYA and TAGAYA, 1972). In each instance, the HA⁻ trait is correlated with the capacity to cause cell-to-cell fusion or polykaryocytosis, identifying these mutants as fusion positive (F⁺). In some exceptional cases, as with HA⁺ and F⁺ cowpox, formation of small polykaryocytes may occur, because the HA is produced more gradually than in the case of H⁺F⁻ vaccinia virus (ICHIHASHI and DALES, 1971).

Table 8. *Effect of rifampicin and hydroxyurea (HU) on production of hemagglutinin and polykaryocytosis in singly and mixedly infected cells* [a]

Virus type	Controls		Rifampicin[b] added		HU added	
	Fusion	HA formation	Fusion	HA formation	Fusion	HA formation
Vaccinia IHD-W	+	−	−	−	−	−
Vaccinia IHD-J	−	+	−	+	−	−
Vaccinia NR₄[c]	+	−	+	−	−	−
Cowpox CP 58	+	+	−	+	−	−
IHD-W + IHD-J	−	+	−	+	−	−
NR₄ + IHD-J	−	+	−	+	−	−

[a] From ICHIHASHI et al., 1971.
[b] Rifampicin was present at 80 μg/ml and HU at 10^{-5} M.
[c] Rifampicin resistant mutant of IHD-W.

Syncytiogenesis is related to intercellular spread of progeny and is contingent upon migration of mature virions to the cell surface (ICHIHASHI et al., 1971). Since expression of both HA and fusing capacity depends on whether glycosylation of virus-specified, membrane-associated polypeptides occurs (WEINTRAUB and DALES, 1974; WEINTRAUB et al., 1977; PAYNE, 1979), it has been suggested that the HA⁻ mutants induce synthesis of a polypeptide moiety which does not become glycosylated (WEINTRAUB and DALES, 1974). In dual infections with HA⁺ and HA⁻ poxviruses, the HA⁺ phenotype is dominant over the F⁺ phenotype (Table 8) (ICHIHASHI and DALES, 1971), in line with the view that the presence of HA and polykaryocytosis are mutually exclusive. This notion is supported by experiments with inhibitors of glycosylation which simultaneously block HA synthesis and allow some syncytiogenesis in infections with HA⁺, F⁺ vaccinia virus. Therefore, a frequently arising spontaneous genetic defect related to HA synthesis can profoundly influence the mechanism for spread of the progeny. This implies that cell to cell dissemination with the HA⁻ strains, which takes place in cell culture despite the presence of neutralizing antibody in the overlay medium (NISHMI and KELLER, 1962), might also occur in animals despite induction of humoral immunity, permitting an intercellular spread of infection both within and between different tissues and organs of the body. By contrast, when HA is present on the wrapping membranes of virions or on membranes of cells containing trapped progeny (ICHIHASHI *et al.*, 1971; PAYNE, 1979), neutralization and disposal of infectious particles could be subject to humoral or cellular immune surveillance. For example, attachment and complexing of HA-positive membranes with circulating erythrocytes might facilitate elimination of infected cell-virus complexes by means of the phagocytes.

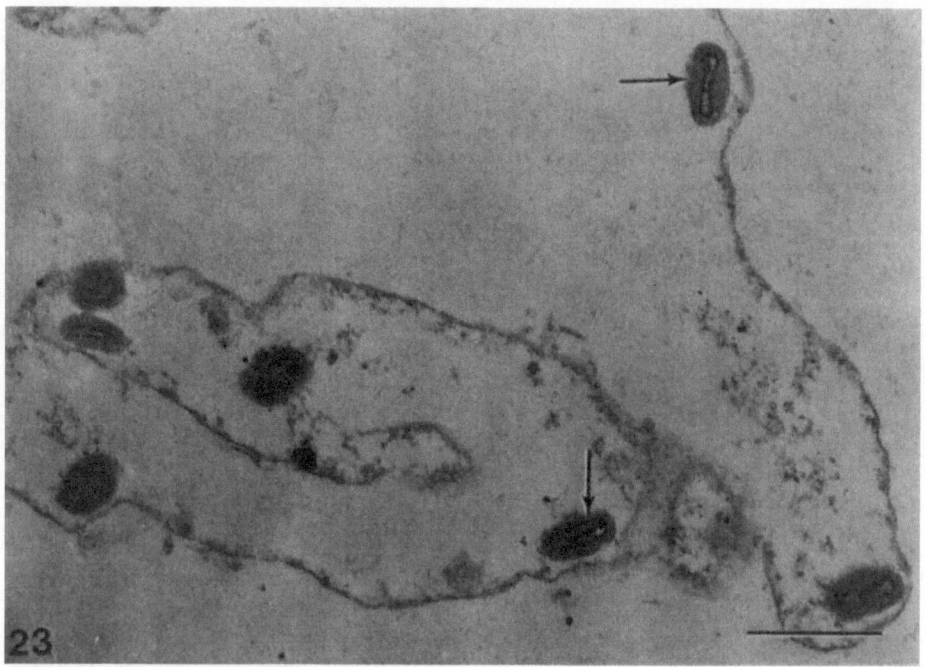

Fig. 23. Selected area of a thin section made from a pellet of purified plasma membranes isolated from HeLa cells 18 hours after inoculation with vaccinia virus. Mature virions attached to the inner face of the membrane are indicated by arrows. × 39,000. (WEINTRAUB and DALES, 1974)

B. Occlusion in Proteinaceous Capsules

Formation of the acidophilic dense cytoplasmic inclusions termed currently "A"-type, which were named initially after their discoverers as BOLLINGER (1873), MARCHAL (1930), or DOWNIE (1939) bodies, is evident during replication of various strains of cowpox, ectromelia and fowlpox viruses. Such inclusions are absent after infection with other vertebrate poxviruses (KATO and KAMAHORA, 1962). These inclusions have been termed "A"-type by KATO and colleagues to distinguish them from Guarnieri bodies, synonymous with the term factories adopted here, which were designated as "B"-type inclusions. Indeed, comprehensive cytochemical studies (KATO and KAMAHORA, 1962; KATO et al., 1962a, b), revealed the proteinaceous composition of "A"-type inclusions (ATI) and showed convincingly that they are distinctive from the factories. The role of ATI in dissemination of progeny virus, suggested by studies with the light microscope (KATO et al., 1962a, b), became clear following the electron microscopic demonstration by MATSUMOTO (1956), and ICHIHASHI and MATSUMOTO (1968a), of occluded mature virions within the proteinaceous matrix. Subsequent investigations from the same and other laboratories revealed that variants of cowpox and ectromelia viruses can be selected which, although they induce ATI, fail to cause virion occlusion (ICHIHASHI et al., 1971; KATO et al., 1963a; ICHIHASHI and MATSUMOTO, 1968a, b). One prerequisite for occlusion is the ability of progeny virions to migrate out of the factories. However, assembly into mature virions is not obligatory, as evident from the finding that cowpox virions arrested at the spherical, immature stage of development can become occluded in ATI (ICHIHASHI and DALES, 1971). The factor for occlusion (V0) is present on the surface of virions (ICHIHASHI and MATSUMOTO, 1968b) and is integrated in virus particles from a pool of soluble material a short time before occlusion takes place (SHIDA et al., 1977). Cowpox strains which can become occluded are designated as V+, while those which cannot, as V−. In dual infections, the presence of the dominant V+ strain provides the V0 factor for integrating V− virions into the ATI (ICHIHASHI and MATSUMOTO, 1968a; ICHIHASHI and DALES, 1971). Related strains of poxviruses, including the IHD-strains of vaccinia virus, which themselves cannot induce ATI, are never-theless, V+, express the V0 factor, and rescue the deficiency for occlusion of V− cowpox strains (ICHIHASHI and MATSUMOTO, 1968a, b).

The unusual translation complexes that carry out rapid synthesis of large pools of the single species of ATI protein comprise numerous, very long polyribosomes, which are somehow bound to the periphery of the ATI (Fig. 24A, B). Using a combination of biochemical, immunological and cytological procedures, in conjunction with specific inhibitors of RNA and protein synthesis, it was estab-lished that (1) ATI are expressed as a late-late cowpox virus function, (2) the associated polyribosomes contain mRNA transcribed from genome sequences specific to cowpox virus and absent from vaccinia virus, (3) the length of poly-ribosomes is appropriate for synthesis of the ATI polypeptide of MW 200K according to recent analysis (SHIDA and MATSUMOTO, personal communication), (4) the polyribosomes are engaged in translation, and (5) occlusion of mature virions is progressive during growth of the ATI (Fig. 24) (ICHIHASHI and DALES, 1973).

Although evidence for the role of ATI in maintaining the viability of vertebrate poxviruses and their spread between animal hosts is only circumstantial, experi-mental data regarding the function of similar inclusions in the survival and dissemination of insect agents is more direct. Inclusions synthesized under the control of insect poxviruses have been termed spherules because of their quasi-spherical shapes (BERGOIN et al., 1971). They are constituted from protein(s) which are organized to form an orderly crystalline lattice (Fig. 25) (BERGOIN and DALES, 1971; GRANADOS and ROBERTS, 1970; BERGOIN et al., 1971; BERGOIN et al., 1969; STOLTZ and SUMMERS, 1972; ROBERTS and GRANADOS, 1968; MCCARTHY et al., 1974; MILNER and BEATON, 1979). Spherules which usually

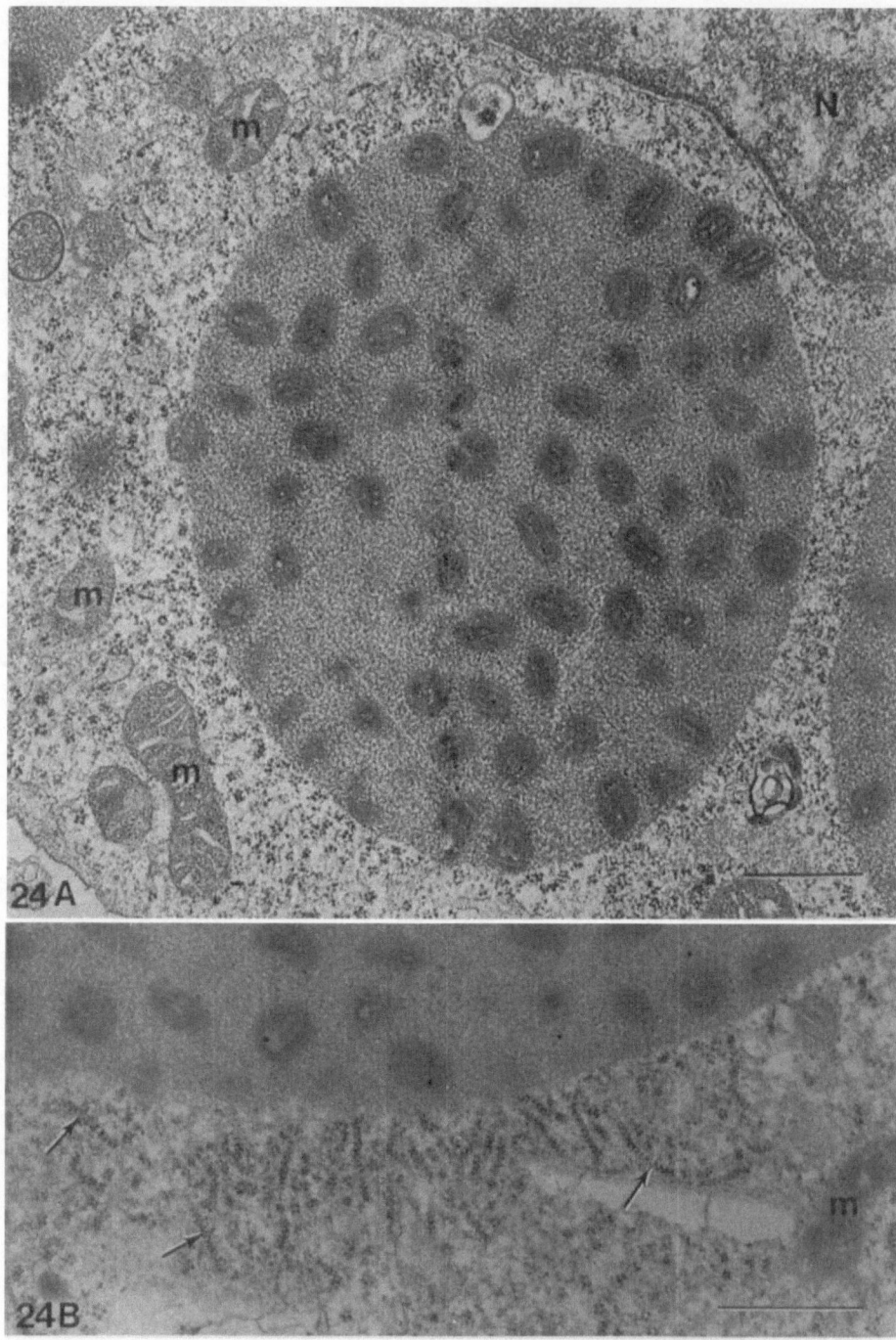

Fig. 24. Selected areas of a HeLa cell sampled 24 hours after inoculation with cowpox virus strain CP 58. *A* The dense A-type inclusion occupying the center of the field is permeated by mature progeny virions. *B* The edge of an inclusion shown to illustrate encrustation by numerous, very long polyribosomes (arrows). Available evidence suggests that such attached polyribosomes (arrows) are engaged in translation of mRNA for A-type protein. *N* nucleus; *m* mitochondrion. *A* ×35,000; *B* ×44,000.
(From Ichihashi *et al.*, 1971)

number one or two per cell appear to develop by accretion of protein and simultaneous integration of virions. Occasionally, individual spherules may attain a diameter of almost 25 μm and enclose several hundred virions. The spherules are liberated into the soil following death and decomposition of infected larvae, thereby providing reservoirs of the agent for spread of the infection (BERGOIN et al., 1971). The incidence of infection may reach 100% in mature larvae of some insect populations (MILNER and BEATON, 1979), testifying to the protective function of the crystalline inclusions for maintaining viability of the virus and facilitating its spread. Upon reinfection of a new host, after ingestion of spherules,

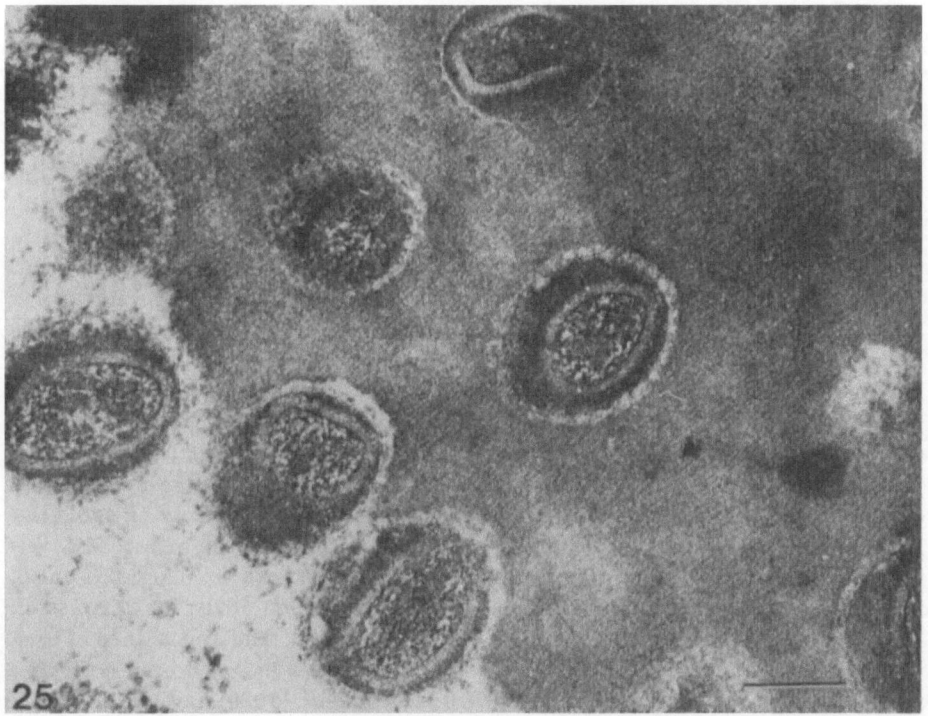

Fig. 25. Thin section of a developing spherule or A-type inclusion of *Melolontha* poxvirus. The mature virions are in the process of being occluded within a proteinaceous, crystalline matrix. × 77,000. (From BERGOIN et al., 1969b)

the occluded virions are freed within the larval gut by dissolution of the crystalline matrix, which is favoured by the prevalence of a high pH in the gut lumen. The liberated virions may then pass through the gut epithelium and invade the body cavity. *In vitro* solubilization of spherules at high pH may be due to activation of a latent protease in them (BILIMORIA and ARIF, 1979). Under these conditions, there is a very rapid release of occluded virions, simulating the *in vivo* phenomenon which takes place in the alkaline environment of the intestinal tract of susceptible insect hosts (MCCARTHY et al., 1974; MILNER and BEATON, 1979).

VIII. Pathology and Disease

A. Events at the Cell Surface

With the notable exception of the agent of *Molluscum contagiosum* (MC), which fails to replicate in any *in vitro* system of primate and nonprimate mammalian cells so far tested, the poxviruses of warm blooded vertebrates generally cross species barriers, infecting a broad spectrum of avian and mammalian cell cultures (FENNER *et al.*, 1974; CHO and WENNER, 1973). It is not surprising, therefore, to find that several mammals, including rabbits, mice and the African rodent *Mastomys natalensis*, are infectible with monkeypox virus (MARENNIKOVA and SELUHINA, 1976; KITAMURA and OGATA, 1979). Accidental infections of man and animals in zoological gardens, including cheetahs (*Acinonyx jubatus*) and other carnivores of the family *Felidae* and elephants with cowpox or cowpox-like agent(s) (BAXBY *et al.*, 1979; MARENNIKOVA *et al.*, 1977; BAXBY and GHABOOSI, 1977), and of rhinoceroses with fowlpox virus (MAYR and MAHNEL, 1970), have been documented recently.

The *in vitro* cytopathology (CPE), which develops as a result of these infections is generally quite uniform, evolving from an early cell rounding (APPLEYARD *et al.*, 1962; BABLANIAN *et al.*, 1978b), to the appearance of massive cytoplasmic inclusions ("factories"), followed by the development of extensive granularity, loss of cell membrane integrity and, finally by cell death and detachment. The interval to cell killing is highly variable, depending on the type of poxvirus and the host cell infected, but it can be as rapid as 10 to 12 hours in the case of virulent strains of variola and vaccinia viruses or killing may occur after several days, as is the case with avian pox, Yaba and rabbit fibroma viruses, which infect the epidermis (CHO and WENNER, 1973). Early cell rounding, evident within 30 to 60 minutes postinoculation, has been correlated with virus-specified protein synthesis, as evident from the application of appropriate metabolic inhibitors (BABLANIAN, 1968; BABLANIAN *et al.*, 1978a; APPLEYARD *et al.*, 1962). In the case of MC, only the early genome functions originating from the coated core are expressed during the *in vitro* infection (SHAND *et al.*, 1976; McFADDEN *et al.*, 1979; LA PLACA *et al.*, 1967; BARBANTI-BRODANO *et al.*, 1974; LA PLACA, 1966). The appearance of early virus surface antigen(s) induced by different poxviruses has also been correlated with cell rounding (McFADDEN *et al.*, 1979; UEDA *et al.*, 1969) and an increase in cell volume and a proportional decrease in cell density, as ascertained by centrifugation through Ficoll gradients (BALL and MEDZON, 1973). The role of such antigens in early CPE is substantiated by mutants of cowpox, variola, monkey pox and vaccinia viruses which are unable to cause cell rounding and which are defective for early antigen production (ITO and BARRON, 1972; AMANO *et al.*, 1979). The early CPE has also been ascribed to soluble cytotoxic factor(s) produced as late functions in the infectious cycle (WOLSTENHOLME *et al.*, 1977), and to leakage of hydrolases from lysosomes (SCHÜMPERLI *et al.*, 1978).

Among the polypeptides implicated in early cytopathology is the component of surface tubular elements (BURGOYNE and STEPHEN, 1979; MBUY and BUBEL, 1978) originally isolated and characterized by STERN and DALES (1976b). STE were shown to enhance cell agglutinability by concanavalin A in the same manner as inoculation with whole virus, implying that, upon normal interaction with the cell membrane, the STE migrate from the inoculum virions and become widely dispersed within the plane of the membrane, accounting for early changes in cell form and increase in agglutinability (MBUY and BUBEL, 1978). During productive infection or when more extensive expression of virus functions occurs, the early cell-rounding phenomenon leads to irreversible and profound disturbances in host cell structure and funtion. By contrast, the abortive infection of primate cells with MC virus *in vitro* is characterized by only a transitory cell rounding. This CPE is reversed in 1 to 3 days, at the time the early surface antigen(s) becomes dissipated and the host cells resume their normal morphology (McFADDEN *et al.*, 1979).

Among several late antigens which appear at the cell surface is the nonvirion hemagglutinin (HA) of vaccinia, cowpox and other orthopoxviruses which is expressed as a late-late dominant (i.e. HA^+) function (ICHIHASHI and DALES, 1971). The spontaneous appearance in infected animals of vaccinia virus mutants which fail to induce HA (i.e. are HA^-) has been correlated with both an absence of glycosylation of the HA polypeptide (WEINTRAUB and DALES, 1974), and acquisition of the capacity to induce cell-cell fusion. Syncytiogenesis appears to require the movement towards and eventual presence at the host cell surface of mature virus progeny (ICHIHASHI and DALES, 1974). The fact that specific antibody to STE suppresses syncytiogenesis (STERN and DALES, 1976b), implies that a component of the STE may control this intercellular fusion process and thereby permit a direct intercellular spread of virus, obviating the usual extracellular phase. Furthermore, one might speculate that the outcome of the disease *in vivo* might be profoundly influenced if infected leukocytes, presumed to be involved in the systemic spread of vaccinia infection, were HA^- at their surface, whereby agglutination with circulating erythrocytes would fail to occur (WEINTRAUB and DALES, 1974). This speculation is based on the reported weak hemagglutination of some mammalian erythrocytes, including human erythrocytes, by vaccinia virus infected cells (CLARK and NAGLER, 1943). Another late antigen is manifested at the cell surface in benign tumors of rabbits inoculated with Shope virus; this antigen elicits a specific antibody response (TOMPKINS *et al.*, 1970c). It is noteworthy that introduction of both early and late poxvirus antigens into the plasma membrane is not accompanied by a displacement of preexisting host proteins and glycoproteins, or by an obliteration of certain cellular biological functions, among them the Na^+- and K^+-dependent ATPase and polio virus receptors (WEINTRAUB and DALES, 1974). Apart from the appearance during infection of glycosylated and nonglycosylated poxvirus-specified proteins at the plasma membrane, other early and late polypeptides, some of which are modified by glycosylation and sulfation, accumulate in the extracellular *milieu* (McCRAE and PENNINGTON, 1978). The function, if any, of such secreted molecules in relation to the infection remains obscure at this time.

No specific cause has been ascertained to explain the profound disorganization of cellular architecture, loss of control over permeability and ultimate cell death observed during the later stages of infection. However, suggestive evidence has been provided implicating leakage of lysosomal enzymes, due to membrane fragility ensuing from infection by Shope fibroma or rabbit poxvirus, with lethality. A decrease of the intracellular activities of specific lysosomal and nonlysosomal enzymes, among them lactic dehydrogenase, acid phosphatase and β glucuronidase, is concomitant with an extracellular increase of the same enzymes within 10 hours postinoculation (SCHÜMPERLI et al., 1978; OGIER, 1974), implying that pathogenesis is related to enzyme leakage.

B. Hyperplastic Response

Those poxviruses which replicate primarily in the epidermal layers of skin, such as fowlpox virus (CHEEVERS et al., 1968), Yaba virus associated with benign histiocytomas in primates (MILO and YOHN, 1975), MC virus of man (VREESWIJK et al., 1977; VREESWIJK et al., 1976, LA PLACA et al., 1967), and Shope rabbit fibroma virus (ANDREWES and AHLSTROM, 1938; FENNER et al., 1974), all may stimulate cell proliferation and hyperplasia rather than causing rapid cell destruction of the type characteristic of vaccinia virus infection. Consequently, nodules or benign tumor-like growths may appear which are usually self limiting in size, because, with time, an immune response is evoked and virus dissemination is suppressed. In some individuals, the MC nodules become widely disseminated to all areas of the body except the palms of the hands and the soles of the feet. These nodules may fail to regress for prolonged periods of time, perhaps because of a deficient immune response to this agent. The occasional appearance of MC lesions in the vicinity of female genitalia suggests that a venereal transmission of the disease may also occur (WILKIN, 1977). In vitro inoculation of susceptible cells in monolayer culture with MC, Yaba or fibroma viruses may result in a loss of contact inhibition coincident with the appearance of foci of piled up cells termed "micro tumors" (SCHWARTZ and DALES, 1971; BARBANTI-BRODANO et al., 1974; TOMPKINS et al., 1969; MILO and YOHN, 1975; YOHN et al., 1970; CROUCH and HINZE, 1977), simulating the in vivo pathological picture evident in the epidermis (PROSE et al., 1969). Although the poxviruses per se are not known to cause malignant transformation, they may act as co-carcinogens. This phenomenon has been demonstrated by inoculating vaccinia virus into mice the skin of which had been painted with methylcholanthrene (DURAN-REYNALS and STANLEY, 1961), or by inoculating Shope fibroma virus into rabbits simultaneously painted with coal tar (ANDREWES and AHLSTROM, 1938).

C. Metabolic Derangements

Infection by poxviruses usually results in profound dysfunction of host cell metabolism. These disturbances may occur as a consequence of the introduction of specific proteins as components of the inoculum, as exemplified by a rapid inhibition of cell DNA replication produced by active and UV-killed vaccinia virus (KATO et al., 1962b, 1943b; MAGEE et al., 1960; KIT and DUBBS, 1963; KIT and DUBBS, 1962a; POGO and DALES, 1973; POGO and DALES, 1974; MOSS, 1974), or Shope fibroma (CHAN and HODES, 1973) and MC viruses. This effect may be due to hydrolysis of the nascent, rapidly labelled 4s short cellular ssDNA fragments by the pH 7.8 nuclease released in the cytoplasmic matrix from inoculum virus cores, which migrates into the nuclear compartment (OLGIATI et al., 1976; POGO and DALES, 1973; POGO and DALES, 1974). By contrast, the preformed host dsDNA is essentially unaffected (OLGIATI et al., 1976; POGO and DALES, 1974; KIT and DUBBS, 1962b). When poxviruses initially elicit proliferative responses in the epidermis, as in the case of fowlpox, Shope fibroma and Yaba viruses, host nuclear

DNA synthesis is actually stimulated prior to the onset of cytoplasmic virus-related DNA replication (CHEEVERS et al., 1968; TOMPKINS et al., 1969). With epidermal MC infection, commencement of virus-related cytoplasmic DNA replication coincides with the onset of a decline in host DNA synthesis (TANIGAKI and KATO, 1967).

Unlike the case with some DNA bacteriophages, the host DNA in poxvirus-infected cells is not converted to reutilizable acid soluble products (SHEEK and MAGEE, 1961). More recently, PARKHURST et al. (1973), have clearly shown that although host DNA does not become a substrate for the synthesis of the vaccinia virus genome, limited DNA breakage occurs within 90 minutes postinfection. This cleavage is dependent upon the multiplicity of the inoculum employed. These observations support the evidence of POGO and DALES (1973, 1974), who demonstrated migration of the core pH 7.8 endonuclease from inoculum virions into nuclei. Coincidentally, host nuclear DNA polymerase activity, assayed in vitro, was found to be inhibited.

Inhibition of host-specified RNA synthesis occurs less rapidly (KIT and DUBBS, 1962a), beginning with mRNA within 3 hours and causing an almost complete cessation of all cellular RNA formation by 7 hours postinfection (BECKER and JOKLIK, 1964). It has been suggested that the decline in host RNA synthesis may be functionally related to a profound decrease of the enzyme uridine kinase, which is observed following vaccinia virus infection (KIT et al., 1964). However, the actual mechanisms responsible for cessation of host RNA formation have not as yet been satisfactorily elucidated.

Suppression of RNA formation is evident between 4 to 6 hours into the vaccinia virus replication cycle (BECKER and JOKLIK, 1964). Transport of mRNA into the cytoplasm and processing of ribosomal RNA precursors is blocked 2 to 3 hours after infection. Unlike the inhibition of host DNA and protein formation, suppression of host RNA production requires synthesis of virus-specific DNA and protein (BARLEY, POGO and DALES, unpublished results). A possible target is the host aggregate RNA polymerase activity, perhaps affected by the accumulation of a basic viral protein. An unpublished study (BARLEY, POGO and DALES) of the nuclear DNA-dependent RNA polymerases was conducted, taking advantage of the possibility of testing nuclear enzymes in situ by means of activating cations and specific inhibitors. The data indicate that, at 4 hours postinfection, α-amanitin-sensitive host RNA polymerase II becomes inhibited by 50%, and later, at 8 hours postinfection, activity of RNA polymerase I is also suppressed. The latter effect may be related to a block of host protein formation because polymerase I activity is more sensitive to disruption of protein synthesis. The decrease in RNA polymerase II activity at the earlier time point is unlikely to be due to a soluble inhibitor appearing after infection, as shown by mixing nuclei from control and infected cells. It remains possible that RNA polymerase II leaks out of infected nuclei, and is engaged in transcription of vaccinia virus DNA in the cytoplasm, because our findings demonstrated that this host cell enzyme is required to transcribe genes involved in late stages of vaccinia virus maturation (SILVER et al., 1979).

Numerous studies have shown that repression of host-specified protein synthesis occurs as rapidly as the inhibition of cellular DNA synthesis. These effects are initiated within 20 minutes after inoculation with vaccinia virus, and are almost complete within 1 to 4 hours, depending upon the multiplicity of infection (Moss, 1968). The roles of virus-specified transcription and translation in this repression are as yet unclear, and evidence has been presented implicating either translation (DRILLIEN et al., 1978), or transcription (BABLANIAN et al., 1978a; KIT and DUBBS, 1962a). Contradictory evidence has, however, also been provided to show that vaccinia virus cores interrupt translation directly in an in vitro system (BEN-HAMIDA and BEAUD, 1978), and inoculum vaccinia made nonfunctional by ultra-violet irradiation is capable by itself of shutting off protein synthesis (Moss, 1968). This has been ascribed to the STE component, since purified STE alone can rapidly and specifically block host protein synthesis, presumably by affecting the initiation of translation directly, as indicated also by STE effects on the reti-culocyte cell-free system (MBUY et al., personal communication). Another recent study suggests the possibility that initiation of translation is influenced by some early transcription product of vaccinia virus (SCHROM and BABLANIAN, 1979), since the short poly(A) chains synthesized under the direction of the virus core enzyme poly(A)-polymerase are able to compete with and displace host mRNA engaged in in vitro translation (ROSEMOND-HORNBEAK and MOSS, 1975).

Despite the suppression of host protein synthesis, the formation of arginine by anabolism of citrulline is enhanced after infection, presumably to fulfill the high requirement for arginine in rabbitpox virus and vaccinia virus replication (OBERT et al., 1971; COOKE and WILLIAMSON, 1973). Some evidence is at hand that such enhanced arginine synthesis may, in fact, be under viral rather than host direction (WILLIAMSON and COOKE, 1973; HODGSON and WILLIAMSON, 1975).

Although infection with various poxviruses generally causes an overall inhibi-tion of host metabolism, there may exist an obligatory requirement for selected synthetic activities of the host cell, because inoculation of human peripheral blood leukocytes with vaccinia virus does not result in a productive virus cycle unless these cells are also transformed (i.e., induced into DNA synthesis and transcrip-tion) by a mitogenic inducer such as phytohemagglutinin (MILLER and ENDERS, 1968).

In the short term, suppression of host-related DNA, RNA and protein syntheses resulting from infection by a poxvirus does not appear to influence significantly either the synthesis or composition of cellular phospholipids (STERN and DALES, 1974; GAUSH and YOUNGNER, 1963), but the glycolipids are altered profoundly. The least complex of these, the ceramide monohexoside, becomes markedly elevated, whereas there are concomitant reductions in the amounts of the more complex ceramide trihexoside and the predominant gangliosides (ANDERSON and DALES, 1978). Such changes are most probably the consequence of virus-mediated inhibition of host protein synthesis, because addition of protein synthesis inhibitors to uninfected cells alters glycolipid composition in a similar manner (ANDERSON and DALES, 1978).

D. Systemic Infections

The most detailed earlier studies on poxvirus dissemination throughout an animal, resulting in a generalized infection, were carried out by MIMS (1959, 1964), on mice inoculated with ectromelia or vaccinia viruses. This series of investigations revealed that when massive inocula are injected intravenously, virus particles are cleared from the blood stream within a few minutes by macrophages lining the liver sinusoids. Evidently, not all of the inoculum is eliminated by the macrophages, because residual virus may be found associated with circulating platelets and leukocytes. MIMS, employing immunofluorescence labelling for light microscopy, showed that replication can be initiated within macrophages of both mice and rabbits, despite the role this cell type plays in the primary defense against infecting pathogens. In the case of myxoma virus, a poxvirus virulent for rabbits, spread from liver sinusoids to parenchymal cells takes place during more advanced stages of the disease. Parallel observations were made on ectromelia and vaccinia virus infections in mice. When smaller virus inocula are administered to suckling mice *via* the intracerebral route, replication occurs in cells of the upper respiratory tract, but fails to be manifested in liver parenchymal cells.

By contrast with the fate of inocula containing infectious virus, killed poxvirus particles injected into mice can be shown, by means of the immunofluorescence procedure, to undergo destruction following their uptake into macrophages. Data from *in vivo* studies are closely analogous to the results of *in vitro* experiments which demonstrate that heat inactivated and antibody-neutralized inoculum vaccinia virus is shunted into lysosomes of L cells, where virions are dismembered (Table 7; Fig. 15) (DALES and KAJIOKA, 1964).

The lymphatic system is also involved in the infectious process, as is evident following subcutaneous or intravenous inoculation of mice with ectromelia virus. Upon reaching the spleen, this virus is taken up by macrophages lining the splenic sinusoids, commencing replication in a few cells within 7 hours after inoculation (MIMS, 1959; 1964). Virus may also be introduced into the spleen and other lymphatic tissue *via* circulating, infected lymphoid cells.

If an active process of virus replication is initiated in the spleen, evidence of extensive destruction of splenic follicles can be observed within 24 hours. The interval required for the appearance of ectromelia virus in local lymph nodes is related to the size of the inoculum, being only a few minutes after subcutaneous injection of a large virus dose but occurring much later, at about 24 hours, when only 100 PFU are injected, implying that in the latter situation virus is carried into the lymph nodes by circulating cells originating from another site of infection. Infection arising from exposure of mice to an aerosol containing ectromelia virus becomes localized in pulmonary lymph nodes by the 3rd day after infection. This finding is in line with the assumption that poxviruses, whether acquired by the intranasal or subcutaneous routes, gain access indirectly to the vascular system and hence the visceral organs via the lymphatic system. However, it is possible that virus which is propagated initially at the site of injection in the epidermal or subcutaneous tissue can also enter directly into the blood without necessarily having to pass first through the lymph nodes.

A poxvirus inoculum introduced into the peritoneal cavity is able to replicate in both the free macrophages and those attached to mesenteries and it also infects lymphocytes situated in this body compartment. The intraperitoneal route may cause the most lethal disease produced by ectromelia virus in mice, because it provides direct access to organs located in the abdomen, consequently compromising the immune response.

Concerning invasion of connective tissue by poxviruses, the macrophage is the only available cell type of the reticuloendothelial system capable of disposing of the inoculum and it may be the cell which initially becomes infected after subcutaneous inoculation of mice with ectromelia and vaccinia viruses and rabbits with myxoma virus. Presumably, further virus spread and development of a generalized infection might be arrested at this stage if macrophages of the subcutaneous connective tissue, "primed" by a previous exposure to the infecting agent, are able to eliminate the inoculum virions completely.

Aerosol-borne infection of mice with ectromelia virus is likely to become localized in the lung, where macrophages and susceptible alveolar cells of the mucosa are the targets. In this location, replication may occur in the mucosal cells, but macrophages carrying the sequestered inoculum are likely to transfer it to the mucociliary region in the lower respiratory tract, from where the virus can enter the pulmonary lymph nodes and, perhaps, also the spleen. It thus appears that when the respiratory tract is the portal of entry, a vascular spread of the infection may initially involve alveolar macrophages.

Intracerebral inoculation with vaccinia, ectromelia or rabbitpox viruses leads to virus replication in cells lining the cerebrospinal fluid spaces, causing meningitis and ependymitis (MIMS, 1964). There is no clear-cut evidence that these agents can infect the neurons (MIMS, 1960). However, dissemination of virus has been traced from the brain to the thoracic and abdominal viscera and a correlation has been established between virus titres in the blood and the extent of mononuclear cell infection (GINSBERG and JOHNSON, 1976). The quantity of free virus recoverable from the organs tested decreases in relation to the potency of neutralizing antibody appearing in the circulation. Of course, dissemination of poxviruses into the central nervous system compartment after establishment of a primary infection elsewhere results in a secondary encephalitis of the type known in man as postvaccinial encephalitis, a rare complication of vaccination against smallpox.

From the above considerations the importance of macrophages in the pathogenesis of poxvirus disease becomes clear. The cells are disposed at surfaces of tissues and compartments in the body where they can control the susceptibility of target organs in an animal to virus infections or the induction of immune responses. Disease and pathology similar to the processes seen in poxvirus infections of mice and rabbits is manifest also when *Cynomolgus* monkeys are injected intramuscularly with monkeypox virus (CHO and WENNER, 1973), an agent highly virulent for this primate. Spread of monkey pox in its host, illustrated in Fig. 26, gives a picture that is closely analogous to smallpox produced by variola virus in man.

Intrathalamic inoculation of vaccinia virus into *Cynomolgus* monkeys causes widespread inflammation in the meninges and choroid plexus which can be directly ascribed to virus replication (MORITA *et al.*, 1977). However, involvement of brain parenchymal tissue (encephalopathy), which occurs during the terminal stages of the disease, is thought to be the consequence of an interruption of blood flow to the CNS rather than a direct effect of monkeypox virus on the brain itself (GHENDON *et al.*, 1973).

The orthopoxviruses are highly variable in neurovirulence. When several variants of the prototype vaccinia virus were tested, it was shown that some strains, including CV-1 and Ikeda, may cause 100% mortality, even in small doses. But other variants, among them a temperature-sensitive variant, LC-16, derived from the Lister vaccine strain and a small pock isolate, VI, originating from the Dairen-I strain, fail to kill test animals even when 100 times greater inocula are injected.

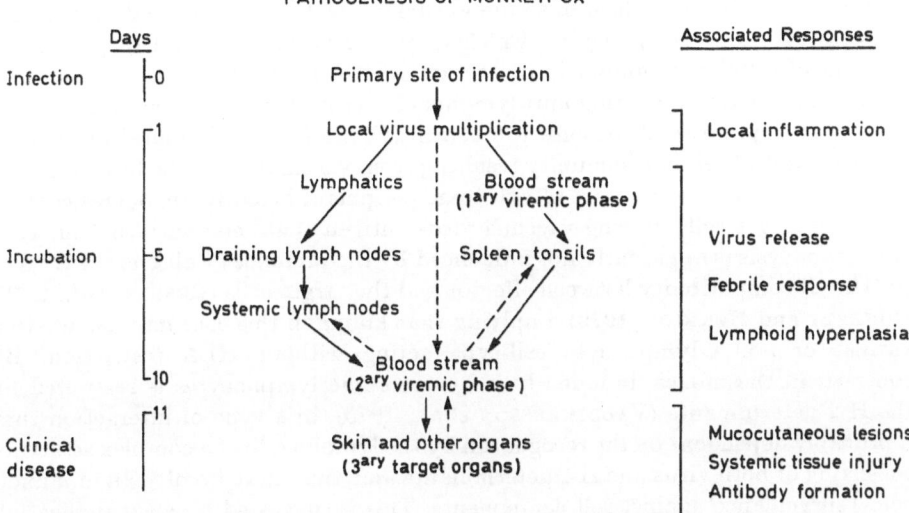

Fig. 26. Model for the pathogenesis of monkeypox. The model is based on data derived from MPV infected intramuscularly in *Cynomolgus* monkeys. (From Cho and Wenner, 1973)

E. Infection and the Immune System

Both humoral and cellular immune responses to poxviruses generally conform to patterns of response evoked by most other infectious agents. Hemagglutination-inhibiting (HI) and neutralizing antibodies are detectable within 8 days after infection of *Cynomolgus* monkeys with monkeypox virus, almost coincident in time with the first appearance of skin eruptions (Cho and Wenner, 1973). These types of antibodies reach peak titers between the 3rd and 6th week and thereafter slowly decline.

Complement-fixing (CF) antibody generally follows the pattern of appearance and decline observed with the above classes of antibody, but its initial formation occurs about one week later. The potency of antibodies made in response to monkeypox virus infection depends to some degree upon the quantity of virus inoculated. The presence of common orthopox virus antigens in monkey pox, variola and vaccinia virions, among them the STE (Stern and Dales, 1976b), is the basis for the ability of vaccinia virus, the least virulent of the three, to serve as an effective immunogen against the other two agents. Vaccinia virus does not, however, confer immunity against unrelated Yaba monkey virus, presumably because this agent does not possess any common antigens (Cho and Wenner, 1973). In the case of another virus, MC, virions associated with epidermal nodules can elicit a lifelong humoral immunity in man manifested by continued presence of agglutinating and neutralizing serum antibodies following the initial infection and nodule development (POSTLETHWAITE, 1970).

It is known that peripheral blood leukocytes isolated from vaccinated individuals can be infected *in vitro* by vaccinia virus (MILLER and ENDERS, 1968), suggesting that following their removal from the animal the leukocytes are left unprotected either by a lack of serum antibodies or because vaccination does not activate cytotoxic lymphocytes, which may also be instrumental in protection. The role of cellular immunity in protection against poxviruses was demonstrated in monkeys by administering anti-lymphocyte serum to immunized animals. The disease then progressed, despite the continued presence of humoral antibodies. Development of cellular immunity resulting from vaccination in the human can be demonstrated by the presence of cytotoxic peripheral blood lymphocytes capable of killing target cells bearing vaccinia virus antigen(s) at their surface. This type of immune lysis is apparently not influenced by the presence or absence of HLA-A or HLA-B compatibility between effector and the target cells (PERRIN *et al.*, 1978; DOHERTY and BENNICK, 1979), implying that killing in this case may be due to a variant or non-T-lymphocyte cell-type acting without HLA restriction. By contrast, in the mouse, immune lysis by cytotoxic lymphocytes is restricted by the H-2 determinants (VALDIMARSSON *et al.*, 1975), in a type of interaction that is probably dependent on the recognition by the lymphocyte of a complex structure consisting of both virus and H-2 determinants and, thus, may involve an immunological surveillance against self components. This is suggested because application of anti-H-2 serum to the virus-induced target cells *in vitro* blocks or reduces their lysis, implying that there exists a topological interrelationship between the H-2 and vaccinia antigens at the target cell surface (DOHERTY and BENNICK, 1979; VALDIMARSSON *et al.*, 1975).

In recognizing the prominence of cellular immunity in control of poxvirus infections, key roles have been assigned to macrophages and lymphocytes (TOMPKINS *et al.*, 1970a; KELLER *et al.*, 1979). Restriction, when it occurs, may operate at the cell surface, since the virus inoculum can become attached to the cell membrane without being internalized (AVILA *et al.*, 1972; TOMPKINS *et al.*, 1970a, b). Upon inoculation into either non-immune mice or rabbits, vaccinia virus or Shope fibroma virus is able to multiply in peritoneal macrophages (TOMPKINS *et al.*, 1970b; KOSZINOWSKI *et al.*, 1975). By contrast, peritoneal macrophages from immune animals are entirely non-permissive (TOMPKINS *et al.*, 1970a, b; TOMPKINS and RAMA RAO, 1978; KOSZINOWSKI *et al.*, 1975), but alveolar macrophages in the rabbit remain susceptible to vaccinia virus. Restriction following immunity does not appear to operate as a universal phenomenon because myxoma virus is able to multiply in macrophages from immune rabbits (TOMPKINS *et al.*, 1970a). In the case of Shope fibroma virus, infection results in the formation of epidermal nodules which appear in about 3 days after inoculation of adult rabbits and regress by the 9th to 21st day, when the immune responses come into play (TOMPKINS and RAMA RAO, 1978). However, if fibromas are induced in newborn rabbits, their growth is not limited for 4 to 5 weeks and complete regression may not be evident until the 8th week, although occasionally animals succumb to the disease before the anti-tumor response becomes effective (TOMPKINS *et al.*, 1973). Not until 15 to

21 days after fibromas appear, do macrophages obtained from rabbits infected at birth become refractory to *in vitro* infection, implying that such delay in resistance may facilitate tumor development. Lymphocytes, which also fulfill essential functions in cellular immunity, likewise exhibit greater anti-tumor cytotoxicity when taken from adult tumor-bearing rabbits than do lymphocytes originating from young tumor-bearing animals inoculated at birth (TOMPKINS *et al.*, 1973).

Sera from rabbits in which fibromas had regressed and sera from young animals bearing growing tumors both contained antibody highly cytotoxic for cultured cells infected *in vitro* with Shope fibroma virus (TOMPKINS and SCHULTZ, 1972). But in such sera, specific antibodies directed against non-virion antigen(s) on tumor cells are only weakly active. These findings suggest that Shope virus-specified antigens which become fully expressed in adult animals are somehow modified or masked during the proliferative phase of tumor formation in young animals, despite the presence of some cytotoxic anti-tumor cell antibodies, accounting for the delay in regression of the fibromas (TOMPKINS and SCHULTZ, 1972).

There exists a voluminous literature dealing with humoral immunity in various models involving poxvirus infection of mammals. Thus, neutralizing antibody appears in serum within one week after intradermal inoculation of either vaccinia, cowpox or monkeypox viruses (SOEKAWA *et al.*, 1977). Similarly, inoculation of adult rabbits with Shope fibroma virus elicits circulating antibody, detectable at first within 7 days, which reaches a peak by the 23rd day and remains elevated until at least the 50th day (SINGH *et al.*, 1972). The antibody classes evident in the rabbit fibroma virus infection include IgM, which peaks by the 13th day, then declines to zero by the 17th day and IgG which accumulates less rapidly, not peaking until the 23rd day, then remaining at high level for several weeks or longer. Some of the cytotoxic antibodies may be formed in response to non-virion functions appearing at the cell surface, such as the HA (BAXBY *et al.*, 1979) and fibroma tumor antigen(s) (SINGH *et al.*, 1972). The virus-neutralizing antibodies may be directed against virion surface components, such as the STE located on "naked" particles (STERN and DALES, 1976b), or non-virion proteins and glycoproteins present on the wrapping membranes (see Figs. 22A—D), (APPLEYARD *et al.*, 1971; BALACHANDRAN *et al.*, 1979), derived from modified vesicles of the host's Golgi apparatus (ICHIHASHI and DALES, 1971). These virion and non-virion antigens must serve as powerful immunogens, because, after convalescence from a natural cowpox infection the anti-HA antibody may persist for more than 6 months and the virus-neutralizing antibody is found in the serum for at least 2 years (BAXBY and OSBORNE, 1979).

IX. Genetics

A. Introduction

The occurrence of both minor and gross mutational changes in the genomes of poxviruses has been linked with alterations in host range or in the disease manifested in either the natural or laboratory setting. The effects of such mutations on the infectious processes are amenable to careful analyses, using procedures to evaluate a variety of phenotypic markers, such as the degree of virulence, thermosensitivity, pock or plaque type, serological relatedness, capacity for complementation or recombination, host restriction and defectiveness in virus-specified metabolic processes or virion assembly. New techniques of exquisite sensitivity, utilizing restriction endonucleases, "Southern-transfer" (COOPER and MOSS, 1978), and R-loop mapping make it possible to pinpoint individual mutations due even to single base alterations (McFADDEN et al., 1980, SCHÜMPERLI et al., 1980) and to identify the outcome of such mutations by characterizing changes in single polypeptides by means of O'FARRELL's two-dimensional PAGE (ESSANI and DALES, 1979; McFADDEN et al., 1980). In the case of the orthopoxviruses, studies to date reveal that spontaneous mutations can frequently arise in the form of deletions which may be relatively minor, involving fewer than 250 base pairs (McFADDEN and DALES, 1979), or deletions can be more extensive, encompassing a loss of several thousand bases (ARCHARD and MACKETT, 1979; MOYER and ROTHER, 1980). Both types of excision occur in the vicinity of the terminal cross-links in the genome and involve the inverted terminal repetition sequences (DE FILIPPES, 1976; MACKETT and ARCHARD, 1979; WITTEK et al., 1978a; GARON et al., 1978).

We shall now consider in some detail our knowledge of the molecular and biological aspects of poxvirus genetics, placing special emphasis on the heritable changes among these agents that emanate from natural animal reservoirs as well as those pertaining to extensively studied laboratory prototypes.

B. Spontaneous Genetic Variability

Numerous investigations of the orthopoxviruses in man and other mammals have revealed the existence of a close interrelationship among them (ESPOSITO et al., 1977b, c), as well as much genetic variability. This variability is seen principally in certain specific and crossreacting antigens of the virion or in certain virus-induced early and late antigens such as the hemagglutinin (HA), which appear at the host cell membranes. Other parameters associated with spontaneous variability among orthopoxviruses, in natural reservoirs or in laboratory-adapted strains, include manifestations of cytopathology, evident as either the plaque- or pock-type formed in terms of degree of virulence and mortality. Even defined vaccinia virus stocks originating from cloned isolates propagated in tissue culture (SHARP and McGUIRE, 1970) have been shown to consist of particles possessing a great deal of variability in structure and infectiousness. This finding should be examined in the light of evidence of genome length heterogeneity among laboratory stocks, which was demonstrated by restriction endonuclease mapping to occur at the inverted terminal repeat segments (WITTEK et al., 1977; WITTEK et al., 1978b). It is quite plausible that differences in the lengths of terminal sequences may arise from non-lethal minor mirror-image deletions, such as those appearing spontaneously in stocks of IHD-W vaccinia virus (McFADDEN and DALES, 1979). Of greater evolutionary significance may be the terminal length differences which are characteristic of each genome type among sero-related variola, vaccinia, cowpox, ectromelia, and rabbit pox group of viruses (MÜLLER et al., 1977; ESPO-

SITO *et al.*, 1978). By contrast with terminal variability, the internal sequences, equivalent to about 75% of the genome, are highly conserved among this group of orthopoxviruses (Fig. 28), implying that evolutionary changes both gross and minor are primarily confined to the inverted terminal repeat segments. However, some homology occurs near the ends, since these agents were also shown to contain a short segment of common or related base sequences 6×10^6 daltons in length at or near the termini containing the cross-linked, inverted terminal repetitions (WITTEK *et al.*, 1977; MACKETT and ARCHARD, 1979; ARCHARD and MACKETT, 1979).

Patterns of polypeptide and antigenic modulation generally reflect the findings of analyses of genome homology. Use of one-dimensional PAGE has shown the existence of many identical and some distinctive polypeptides among orthopoxviruses related to the group of which vaccinia virus is a prototype (ESPOSITO *et al.*, 1977a; HARPER *et al.*, 1979; TURNER and BAXBY, 1979). From the data already at hand (ESSANI and DALES, 1979), it is to be expected that future application of two-dimensional PAGE to studies of polypeptide variability among poxviruses will provide a very sensitive analytical method for detecting and evaluating phenotypic expressions of mutations related to alterations in specific polypeptides. In this regard, a correlation between a base alteration affecting an EcoRI nuclease restriction site in the vaccinia virus genome and a simultaneous alteration in the charge of core-associated polypeptide has already been demonstrated (McFADDEN *et al.*, 1980).

The application of serotyping to numerous orthopoxvirus isolates, whether originating from man or animals or as spontaneous pock and plaque variants appearing in the laboratory, if conducted with carefully selected antisera, has proved extremely useful in rapid screening of related poxviruses and for distinguishing the group specific crossreacting antigens from those specific to each type or subtype. This method of analysis is of particular value to the epidemiologist interested in following the dissemination of human disease in simian and other vectors, because it permits the identification of and distinction between vaccinia, variola and monkeypox viruses by means of diagnostic antisera (GISPEN and BRAND-SAATHOF, 1974; ESPOSITO *et al.*, 1977b; MARENNIKOVA *et al.*, 1978). More elaborate schemes for typing individual isolates from among 18 or more different serotypes have proved invaluable in diagnosis of outbreaks of poxvirus diseases among domesticated animals and wild animals in captivity (MAHNEL, 1974; MARY *et al.*, 1972). These procedures employ not only serology for identification but also characteristics of plaque or pock type, the nature of cutaneous lesions in rabbits and near feather follicles, pathological changes in embryonated eggs, virulence, and other criteria (BAXBY, 1975; MAYR *et al.*, 1972; MAHNEL, 1974).

Correlations between changes in molecular organization of the genome and phenotypic expression in terms of biological functions have provided new insights into the generation of spontaneous variants. Thus, monkey pox, which usually produces an ulcerative (U+) or hemorrhagic red pock on the chick choriallantoic membrane, mutates abruptly at fairly high frequency into a type causing non-hemorrhagic "white" (U) pocks (SAMBROOK *et al.*, 1969; MARENNIKOVA, 1979). Associated with such an abrupt shift in phenotype is the occurrence of major terminal deletions which can be 11% of the length of the DNA molecule (ARCHARD and MACKETT, 1979), once again emphasizing the plasticity of the orthopoxvirus genome at its termini. Other characteristics which may be altered spontaneously, in relation to or independent of the pock type, include virulence (BAXBY, 1975; MAYR et al., 1978; SHELUKHINA et al., 1979), host range (FENNER and SAMBROOK, 1966; SAMBROOK et al., 1965; MAYR et al., 1978), thermosensitivity (SHELUKHINA et al., 1979), the expression of early antigen(s) (UEDA et al., 1969; ITO and BARRON, 1972), and the appearance of hemagglutinin (HA) at the host cell surface (WEINTRAUB and DALES, 1974; ICHIHASHI and DALES, 1971; CHO and WENNER, 1973; DALES et al., 1976; HANAFUSA et al., 1959). Vaccinia virus variants which fail to express HA (HA⁻), presumably reflecting a defect in glycosylation of the HA polypeptide (DALES et al., 1976), have been isolated repeatedly from infected animals, implying that they arise frequently in nature. Laboratory identification of vaccinia virus variants possessing traits deemed appropriate for vaccine strains, such as low virulence, restricted host range and thermosensitivity, has favoured the selection of numerous isolates suitable for use as human vaccines (MAYR et al., 1978).

C. Induced Genetic Variability

As might be expected from the similar sizes of their genomes, the poxviruses are closely analogous in terms of their biological complexity to the tailed T-even bacteriophages. It is, therefore, not surprising that, as a consequence of recent inventions of exquisitely sensitive procedures useful for analysing eukaryotic materials, investigators have turned their attention to vaccinia virus as a prototype for systematic genetic analysis of complex animal viruses. The Australian group of virologists, particularly F. FENNER, W. K. JOKLIK and J. SAMBROOK, were the first to describe procedures using chemical mutagenesis with bromodeoxyuridine (BUdR) for inducing and isolating a group of conditional-lethal, temperature-sensitive (ts) mutants of rabbitpox virus (SAMBROOK et al., 1966). The desirable criteria which sister mutants should possess were set out by SAMBROOK and colleagues, and include a) the presence of a unique defect in each isolate, b) absence of double or multiple mutations, c) a random distribution of individual mutations on the genetic map. Complementation and recombination analyses of the 26 ts mutants obtained in Australia fulfilled Sambrook's criteria. Another group of ts mutants induced by several chemical mutagens and selected by defectiveness in plaque-enlargement at the elevated temperature was reported subsequently (CHERNOS et al., 1978). A third group of about 90 ts mutants was isolated by DALES et al. (1978), using a selection procedure patterned after that of SAMBROOK et al. (1966). However, mutagenesis of stock IHD-W vaccinia virus was done in this case with N,-N¹-dimethyl nitrosoguanidine rather than with BUdR. In this group of mutants, phenotypic characterization by means of electron microscopy revealed defects in assembly among the individual isolates. These defects could be grouped into 17 classes according to a scheme of ascending progression of development into mature virions (DALES et al., 1978). These vaccinia virus ts mutants were also shown, using complementation and recombination analyses, to contain single mutations (LAKE et al., 1979; MCFADDEN and DALES, 1980). Substantiation for the occurence of point mutations came from investigating one class of 5 mutants. Although all 5 were phenotypically identical in their manner of mimicking membrane assembly defects produced by the drug rifampicin, each defect must be in a different function, since all were able to complement and recombine with one another. Further proof that, in the entire group of over 90 mutants, each carries an alteration in only a single gene, comes from analyses of mutants belonging to the other 17 classes and the discovery of a specific base alteration at one site (Fig. 12), susceptible to cleavage by EcoRI endonuclease in a mutant designated as *ts* 9251 (McFADDEN *et al.*, 1980; SCHÜM-PERLI *et al.*, 1980). This particular alteration is accompanied by a change in the migration of a 37K core polypeptide detectable by isoelectric focussing. The difference in charge of the protein may be due to a substitution of an amino acid residue.

The recombination frequencies (%rf) of poxviruses are generally rather high. For example, the usual values reported for vaccinia and rabbitpox viruses range from <1% to >50%, depending on the particular mutant pair used in the cross (PADGETT and TOMPKINS, 1968; CHERNOS *et al.*, 1978; M. ENSINGER, personal communication; GHENDON, 1972). Although in one study much lower %rf was reported for one class of vaccinia virus *ts* assembly mutants (LAKE *et al.*, 1978), subsequent, more careful analyses of the same group gave %rf values like those usually obtained by others. Despite an overall greater recombination efficiency recorded in our later experiments, assignments of gene sequence were unchanged (ESSANI and DALES, in preparation). Information already at hand makes it evident that close clustering of mutations on the genetic map does not occur even when the mutations are associated with a functionally related class of defects, such as those pertaining to DNA synthesis (McFADDEN and DALES, 1980), or virion membrane assembly (LAKE *et al.*, 1978).

Apart from articles dealing with systematic isolation and characterization of groups of mutants, sporadic reports have appeared in which single vaccinia virus isolates are characterized. Among them is one conditionally defective in assembly (DRILLIEN *et al.*, 1977), another fails to induce the virus-specified DNA dependent RNA polymerase at elevated temperature (BASILICO and JOKLIK, 1968), and a third is defective for induction of uridine and thymidine kinases (DUBBS and KIT, 1964; KIT *et al.*, 1963a—c). Other mutants have been isolated which are resistant to or even dependent upon specific drugs interfering with assembly and maturation, such as rifampicin (SUBAK SHARPE *et al.*, 1969a, b; Moss *et al.*, 1971; NAGA-YAMA *et al.*, 1970), and isatin-B thiosemicarbazone (IBT) (KATZ *et al.*, 1973a, b; APPLEYARD and WAY, 1966).

Abrupt changes in pock type from the hemorrhagic (u+) to "white" pock (u) trait in rabbitpox (FENNER and SAMBROOK, 1966) have been shown to be accompanied by a non-lethal deletion of a major terminal segment of the genome (MOYER and ROTHER, 1980), and in some cases also by a profound alteration of the virus host range (SAMBROOK *et al.*, 1966). For example, among 17 rabbitpox isolates converted from u+ to u, 6 lost the ability to induce most viral functions in porcine kidney (PK-2a) cells and 3 lost the capacity to replicate in L929 mouse fibroblasts, although such variants can complete the infectious cycle in chick embryo fibroblasts (McCLAIN, 1965). Other host-restricted replication defects have been documented (DRILLIEN *et al.*, 1978).

The relative paucity of useful mutants and the availability of only fragmentary information regarding the mutant phenotypes makes it unlikely that the type of

detailed genetic map constructed by means of recombination analysis for certain
bacteria and bacteriophages will become available in the near future for a poxvirus.

D. Rescue and Reactivation Phenomena

The commonly observed inability of rabbitpox u mutants to complement or
recombine with one another is most probably due to excision of a major terminal
segment which is associated with this phenotype (FENNER and SAMBROOK, 1966;
MOYER and ROTHER, 1980). Another aspect of host range variability relates to the
capacity of one strain of poxvirus to rescue the multiplication of an unrelated one
in a non-permissive host. Efficient rescue in Jinet cells of rabbit pox by Yaba
virus, even after partial U.V. inactivation, is one example. However, using the
same pair of virus types, the rescue of rabbit poxvirus in the PK-15 cell line is
very inefficient (TSUCHIYA and TAGAYA, 1977). The reason for this type of host-
mediated restriction is at present unknown.

In conditions permissive for marker rescue, the dominant rabbitpox RPu+
partner can be reactivated, even following partial U.V. inactivation, after coinfec-
tion with RPu variant. Multiplicity reactivation of U.V.-irradiated aggregates of
RPu+ is also possible (ABEL, 1962a). In both conditions resulting in restoration of
virus infectiousness, efficiency of the rescue process was enhanced when carried
out in cells with small cytoplasmic space, as in the case of chick embryo fibroblasts,
presumably allowing for more frequent interaction between virus genomes. When
cells of the KB line, possessing cytoplasmic volumes ten-fold greater than chick
fibroblasts were used, marker-rescue was less frequent (ABEL, 1962b).

In addition to the usual interactions associated with recombination phenomena,
the poxviruses have the ability to undergo non-genetic reactivation by the
so-called Berry-Dedrick phenomenon (BERRY and DEDRICK, 1936). The original
study demonstrated that rabbits inoculated with heat-inactivated preparations of
myxoma virus would develop the lethal myxomatous disease if they were coinfected with rabbit fibroma virus. Subsequently, the possibility of a similar type
of rescue of one virus strain by another was demonstrated in cell culture (KILHAM
et al., 1958; HANAFUSA et al., 1959; JOKLIK et al., 1960a; FENNER and WOODROOFE,
1960; JOKLIK et al., 1960b; JOKLIK, 1962b), thereby pinpointing the Berry-
Dedrick phenomenon as an intracellular event. In vitro reactivation of vaccinia
virus by cell extracts has also been claimed, but not substantiated (ABEL, 1963).
Whenever the genome of the reactivable virus remains viable, as after protein
denaturation by heating or exposure to concentrated urea solution, the genome of
the rescuing partner may undergo limited inactivation by exposure to U.V. light
or nitrogen mustard without losing the capacity to effect reactivation (JOKLIK
et al., 1960a; FENNER and WOODROOFE, 1960; JOKLIK et al., 1960b). The rescue
process does not require simultaneous infection, as evident when the two partners
are inoculated into rabbits or tissue cultures at intervals 1 to 3 days apart (JOKLIK
et al., 1960a; TSUCHIYA and TAGAYA, 1979). The reactivating capability presumably resides in the virus core. It is, therefore, reasonable to assume that transcription of mRNA that specifies the "uncoating factor" and other critical
functions are provided by the reactivating partner on behalf of the reactivable
virus, the enzymatic proteins of which were denatured.

E. Concerning a Natural Reservoir of Poxviruses: Can Genetic Variability Lead to Reemergence of Smallpox?

Abrupt major changes in the physical organization of the genome among naturally occuring orthopoxviruses, which become evident in altered pock and host range phenotype, have already been discussed. The host can also fundamentally modulate the nature of the disease produced by a given poxvirus strain, as is admirably illustrated in the case of the myxoma-fibroma interconversion involving European and South American rabbits of different species (FENNER, 1959). A fibroma virus infecting its original South American host, *Sylvilagus brasiliensis*, produces a benign self-limiting, cutaneous tumor. By contrast, the same agent, infecting its new European host, *Oryctolagus cuniculus*, causes a generalized lethal myxomatosis (FENNER *et al.*, 1974). Following an adaptation period of the myxomatosis virus in the widely spread population of Australian wild rabbits, which became established successfully from imported European rabbit stock, the agent underwent a selection towards a virus type that exists enzootically and causes the more benign fibromatous disease pattern. Several virus substrains differing in degree of virulence have been isolated from Australian rabbits. The reversion from myxoma to fibroma observed in a wild setting reveals the operation of a natural selection process culminating in the establishment of a less virulent agent benefiting the perpetuation of both the virus and its host (FENNER *et al.*, 1974). Similar examples of natural selection by adaptation, involving defined tissue culture systems, have been documented in the laboratory setting. Thus, repeated passage of a cloned stock of WR vaccinia virus through a restrictive host results in selection of variants which have an extended range of host species (GANGEMI and SHARP, 1976).

From the above, it is not difficult to imagine that a poxvirus producing an inapparent or mild animal disease might, after being subjected to the appropriate selection pressure, become adapted to replication in the human, so as to cause a virulent disease like smallpox. The prime candidate virus for conversion from an animal to a human disease appears to be monkey pox. The plausibility of such an interconversion may become more credible following our review of existing knowledge about animal reservoirs of agents related to variola-virus.

The vigorous, systematic programme of the World Health Organization initiated in 1967 for global eradication of smallpox*, has culminated in the entirely successful interruption by 1976 of man-to-man transmission of the disease, when the last case was documented (ARITA and HENDERSON, 1976; ARITA, 1979). During repeated smallpox epidemics occurring previous to 1976, over 600 variola virus isolates were assembled. In this collection there are substrains encompassing intermediate grades between extremes in virulence from *variola major* at one end to *variola minor* at the other end of the spectrum. Therefore, the individual isolates in this variola virus repertoire exhibit the phenotypes of not only 2 major virus strains, but of a wide variety of subtypes. It is instructive that isolates made from patients were occasionally typed not as variola virus but as vaccinia virus. Presumably, the illness in these individuals was a generalized postvaccination

---

* Progress in smallpox eradication. W.H.O. Chronicle **28**, 359—363 (1974). Smallpox surveillance. Wkly. Epid. Rec. **54**, 137—144 (1979).

infection. In one instance, a 1969 isolate termed "Lenny" virus was characterized as a recombinant between variola and vaccinia viruses (ARITA, 1979; BAXBY, 1977). There are at present no firm data to indicate that a natural pool of variola virus exists in areas of Africa and Asia where frequent smallpox epidemics were located in this century. However, abundant evidence has been obtained indicating the existence of a natural pool of monkeypox virus, particularly in an area of the tropical rain forest of Central West Africa where most of the 36 sporadic cases of human infection have been documented since 1970. Human infection with monkeypox virus simulates symptoms produced by variola virus and, on occasion, may be fatal, but unlike smallpox, monkeypox usually is not transferred between humans. In the two exceptional familial cases involving more than one individual, there may have been a simultaneous but independent infection (ARITA, 1979). Extensive screening of monkeys, rodents and birds indigenous to areas where human infections occur revealed an occasional animal sero-positive for monkeypox infection (KITAMURA and OSATA, 1979). Attempts at deliberate virus isolation from tissues of wild animals proved to be successful, once in the case of a rodent species and another time with kidney tissue from a Chimpanzee captured in Zaire. In other instances, culturing kidney cells of Asian monkeys (BAXBY, 1975; ARITA and HENDERSON, 1976), and a Cynomolgus monkey (CHO and WENNER, 1973), led to the isolation of monkey pox. Some of the above isolates produced the non-hemorrhagic u or "white" type pocks upon inoculation onto chick chorioallantoic membranes, but serological tests confirmed the virus to be *bona-fide* monkeypox virus (BAXBY, 1975; GISPEN and BRAND-SAATHOF, 1974). During an isolated episode, 3 African monkeys which had been held in captivity for 3.5 to 4 years developed a smallpox-like disease but their convalescent sera were positive for monkeypox antibodies. Similar serological findings were reported for 3 individuals convalescencing from human monkeypox disease (GISPEN et al., 1976). The above information taken as a whole reveals that a natural reservoir of monkeypox must exist in certain regions of Asia and Africa and that man is a susceptible host for the virus.

It is now known that both fresh human isolates and laboratory passaged stocks of variola virus induce exclusively "white" pocks on the chorioallantoic membrane, so that the term "white poxvirus" has become applied to this agent. The other orthopoxviruses, including rabbitpox, cowpox, ectromelia and monkeypox viruses usually produce the u+ pocks, but occasionally develop spontaneous mutants which yield u type pocks, referred to appropriately as "white" pock variants. The u variants are serologically indistinguishable from the parental wild type virus (BAXBY, 1975; FENNER, 1979). However, recent reports suggesting that "white" variants originating from a stock of monkeypox virus acquired the markers associated with "white poxvirus" raised the serious possibility that monkeypox virus can be changed by an abrupt mutagenic shift into variola virus (MARRENIKOVA et al., 1978; MARRENIKOVA and SHELUKHINA, 1978; MARRE-NIKOVA et al., 1979). Although this suggestion has by no means been proved rigorously (ZUCKERMAN and RONDLE, 1978), the possibility remains that a variola-like variant of monkeypox virus might arise in nature. With this idea in mind, it is worthwhile to recapitulate the following recent salient information about variability among the poxviruses: a) several members belonging to a group which includes vaccinia, variola, monkeypox, rabbitpox, ectromelia and cowpox viruses are serologically closely related, yet distinguishable antigenically from one another; b) members of this group are readily able to cross species barriers to infect other mammals including man (MARRENIKOVA and SHELUKHINA, 1976; MAYR and MAHNEL, 1970; MARRENIKOVA et al., 1977; BAXBY et al., 1979; BAXBY and GHABOOSI, 1977); c) judging by restriction endonuclease analysis, the genomes of these agents are virtually homologous along 25 to 62% of the length of internal segments (Fig. 27) (ARCHARD and MACKETT, 1979; MACKETT and ARCHARD, 1979), while nucleotide sequence homology of the internal segments may be 73 to 95% (MULLER et al., 1977); d) the greatest variability in genome length occurs at the inverted terminal repeat sequences (MACKETT and ARCHARD, 1979), in the region where either minor spontaneous deletions occur frequently, as shown with vaccinia virus (McFADDEN and DALES, 1979), or where major terminal deletions occur by excision of 3 to 20×10^6 daltons of DNA in the case of cowpox (ARCHARD and MACKETT, 1979), and rabbitpox (MOYER and ROTHE, 1980). Such major deletions are invariably accompanied by a shift from the u^+ to the u "white" pock phenotype; e) different isolates of variola virus all have the shortest genome in this

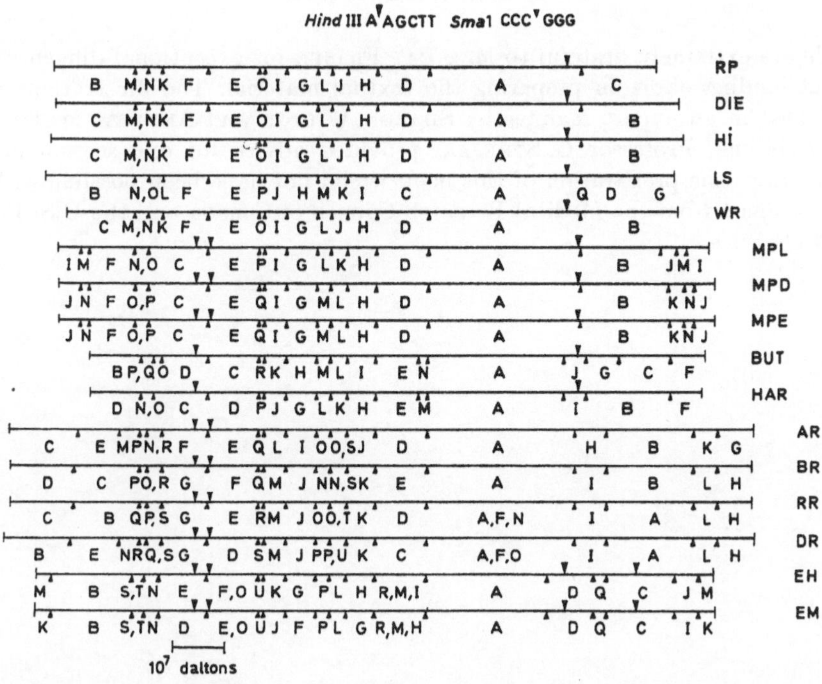

Hind III A'AGCTT Sma1 CCC'GGG

Fig. 27. Physical map locations of Hind III or Sma I restriction fragments of DNA from rabbitpox strain Utrecht (RP); vaccinia strains (DIE), Hall Institute (HI), Lister (LS), or Western Reserve (WR); monkeypox strains Congo (MPC), Denmark (MPD), or España (MPE); variola strains Butler (BUT) or Harvey (HAR), cowpox red strains Austria (AR), Brighton (BR), Ruthin (RR) or Daisy (DR); and ectromelia strains Hampstead (EH) or Moscow (EM). Map locations of Hind III restriction fragments of DNA from rabbitpox or vaccinia strain Lister are from data of WITTEK et al. (1977). (From MACKETT and ARCHARD, 1979)

poxvirus group, presumably as a consequence of having suffered deletion of a segment from one specific end of the chromosome in the region of the terminally repeated sequences (Fig. 27) (MACKETT and ARCHARD, 1979).

It now seems entirely plausible that an agent such as monkeypox virus, endowed with a genome which is inherently plastic at its termini, might, under appropriate selective pressure during multiple replication cycles in a human host, incur non-lethal excision(s), such as those apparent in the DNA of variola virus (MACKETT and ARCHARD, 1979). The spontaneous elimination of a major segment of genome in the emerging variant could then be phenotypically expressed in terms of an increased virulence and altered tropism, facilitating the dissemination of the new virus between humans. It is, therefore, gratifying to report that the W. H. O. authorities, who fully appreciate the existence of a natural poxvirus with a theoretical potential for mutation into a smallpox-like new virus, are carefully monitoring all sporadic cases of human monkey pox and the geographic regions where a reservoir of this virus exists.

Acknowledgments

We are extremely grateful to Mrs. PAT FRASER for exceptional diligence and an outstanding effort in preparing the textual material. The illustrations were assembled in an expert manner by SHARON WILTON and ANDREW MASSALSKI. Our colleague, Professor G. STREJAN, provided advice for one section of the manuscript. The preparation of this book would not have been possible without grant support from the Medical Research Council of Canada and the U.S. Public Health Service.

References

1. ABEL, P.: Topography in vaccinia genetics. Virol. **16**, 347—348 (1962a).
2. ABEL, P.: Multiplicity reactivation and marker rescue with vaccinia virus. Virol. **17**, 511—519 (1962b).
3. ABEL, P.: Reactivation of heated vaccinia virus *in vitro*. Zschr. Vererbungslehre **94**, 249—252 (1963).
4. ALLISON, A. C., VALENTINE, R. C.: Virus particle adsorption II. Adsorption of vaccinia and fowl plaque viruses to cells in suspensions. Biochim. biophys. Acta **40**, 393—399 (1960a).
5. ALLISON, A. C., VALENTINE, R. C.: Virus particle adsorption III. Adsorption of viruses by cell monolayers and effects of some variables on adsorption. Biochim. biophys. Acta **40**, 400—410 (1960b).
6. AMANO, H., UEDA, Y., TAGAYA, I.: Orthopoxvirus strains defective in surface antigen induction. J. gen. Virol. **44**, 265—269 (1979).
7. ANDERSON, R., DALES, S.: Biogenesis of poxviruses: glycolipid metabolism in vaccinia-infected cells. Virol. **84**, 108—117 (1978).
8. ANDREWES, C. H., AHLSTROM, C. G.: A transplantable sarcoma occurring in a rabbit inoculated with tar and infectious fibroma virus. J. Pathol. Bacteriol. **47**, 87—99 (1938).
9. ANDREWES, C., PEREIRA, H. G.: Viruses of vertebrates, 3rd ed., 373—414. Baltimore: Williams & Wilkins 1972.
10. APPLEYARD, G. WESTWOOD, J. C. N., ZWARTOUW, H. T.: The toxic effect of rabbitpox virus in tissue culture. Virol. **18**, 159—169 (1962).
11. APPLEYARD, G., WAY, H. J.: Thiosemicarbazone-resistant rabbitpox virus. Brit. J. exp. Pathol. **47**, 144—151 (1966).
12. APPLEYARD, G., HAPEL, A. J., BOULTER, E. A.: An antigenic difference between intracellular and extracellular rabbitpox virus. J. gen. Virol. **13**, 9—17 (1971).
13. ARCHARD, L. C.: *De novo* synthesis of two classes of DNA induced by vaccinia virus infection of Hela cells. J. gen. Virol. **42**, 223—229 (1979).
14. ARCHARD, L. C., WILLIAMSON, J. D.: The effect of arginine deprivation on the replication of vaccinia virus. J. gen. Virol. **12**, 249—258 (1971).
15. ARCHARD, L. C., MACKETT, M.: Restriction endonuclease analysis of red cowpox virus and its white pock variant. J. gen. Virol. **45**, 51—63 (1979).
16. ARIF, B. M.: Isolation of an entomopoxvirus and characterization of its DNA. Virol **69**, 629—634 (1976)
17. ARITA, I.: Virological evidence for the success of the smallpox eradication programme. Nature **279**, 293—298 (1979).
18. ARITA, I., HENDERSON, D. A.: Monkeypox and whitepox viruses in West and Central Africa. Bull. World Health Organ. **53**, 347—353 (1976).
19. ARMSTRONG, J. A., METZ, D. H., YOUNG, M. R.: The mode of entry of vaccinia virus into L Cells. J. gen. Virol. **21**, 533—537 (1973).
20. ARZOGLOU, P., DRILLIEN, R., KIRN, A.: Evidence for alkaline protease in vaccinia virus. Virol. **95**, 211—214 (1979).
21. ATHERTON, K. T., DARBY, G.: Patterns of transcription of messengers containing poly A in vaccinia virus-infected cells. J. gen. Virol. **22**, 215—244 (1974).

22. AUBERTIN, A. M., McAUSLAN, B. R.: Virus associated nucleases: evidence for endonuclease and exonuclease activity in rabbitpox and vaccinia virus. J. Virol. **9**, 554—556 (1972).

23. AVILA, F. R., SCHULTZ, R. M., TOMPKINS, W. A. F.: Specific macrophage immunity to vaccinia virus: macrophage-virus interaction. Infect. and Immun. **6**, 9—16 (1972).

24. BABLANIAN, R.: The prevention of early vaccinia-virus-induced cytopathic effects by inhibition of protein synthesis. J. gen. Virol. **3**, 51—61 (1968).

25. BABLANIAN, R., ESTEBAN, M., BAXT, B., SONNABEND, J. A.: Studies on the mechanisms of vaccinia virus cytopathic effects. I. Inhibition of protein synthesis in infected cells is associated with virus-induced RNA synthesis. J. gen. Virol. **39**, 391—402 (1978a).

26. BABLANIAN, R., BAXT, B., SONNABEND, J. A., ESTEBAN, M.: Studies on the mechanisms of vaccinia virus cytopathic effects. II. Early cell rounding is associated with virus polypeptide synthesis. J. gen. Virol. **39**, 403—413 (1978b).

27. BAGLIONI, C., LENZ, J. R., MARONEY, P. A., WEBER, L. A.: Effect of double-stranded RNA associated with viral messenger RNA on *in vitro* protein synthesis. Biochem. **17**, 3257—3262 (1978).

28. BALACHANDRAN, N., SETH, P., MOHAPATRA, L. N.: Use of the ^{51}chromium release test to demonstrate antigenic differences between extracellular and intracellular forms of vaccinia virus. J. gen. Virol. **45**, 65—72 (1979).

29. BALL, F. R., MEDZON, E. L.: Sedimentation changes of L cells in a density gradient early after infection with vaccinia virus. J. Virol. **12**, 588—593 (1973).

30. BALL, F. R., MEDZON, E. L.: Evidence for an "early early" vaccinia virus-induced protein which causes a density change of infected L-M cells. J. Virol. **17**, 60—67 (1976).

31. BALTIMORE, D.: Expression of animal virus genomes. Bacteriol. Rev. **35**, 235—241 (1971).

32. BARBANTI-BRODANO, G., PORTONLANI, M., BERNARDINI, A., STIRPE, F., MANNINI-PALENZONA, A., LA PLACA, M.: Thymidine kinase activity in human amnion cell cultures infected with Shope fibroma virus. J. gen. Virol. **3**, 471—474 (1968).

33. BARBANTI-BRODANO, G., MANNINI-PALENZONA, A., VAROLI, O., PORTOLANI, M., LA PLACA, M.: Abortive infection and transformation of human embryonic fibroblasts by *Molluscum contagiosum* virus. J. gen. Virol. **24**, 237—246 (1974).

34. BARBOSA, E., MOSS, B.: mRNA (nucleoside-2′-)-methyltransferase from vaccinia virus. J. biol. Chem. **253**, 7692—7697 (1978).

35. BARBOSA, E., GARON, C. F., MOSS, B.: Transcription of the terminal repetition in vaccinia virus DNA. 2nd Cold Spring Harbor Poxvirus Iridovirus Workshop, 1979. (Abstract.)

36. BARNARD, J. E.: The causative organism in infectious ectromelia. Proc. Roy. Soc. Biol. **109**, 360—380 (1931).

37. BAROUDY, B. M., MOSS, B.: Purification and characterization of DNA-dependent RNA polymerase from vaccinia virus. 2nd Cold Spring Harbor Poxvirus Iridovirus Workshop, 1979. (Abstract.)

38. BASILICO, C., JOKLIK, W. K.: Studies on a temperature-sensitive mutant of vaccinia virus strain W.R. Virol. **36**, 668—677 (1968).

39. BAUER, W. R., RESSNER, E. C., KATES, J., PATZKE, J. U.: A DNA nicking-closing enzyme encapsidated in vaccinia virus: partial purification and properties. Proc. Nat. Acad. Sci. U.S.A. **74**, 1841—1845 (1977).

40. BAXBY, D.: Identification and interrelationships of the variola/vaccinia subgroup of poxviruses. Progr. med. Virol. **19**, 215—246 (1975).

41. BAXBY, D.: Poxvirus hosts and reservoirs. Arch. Virol. **55**, 169—179 (1977a).

42. BAXBY, D.: Possible antigenic sub-divisions within the variola/vaccinia subgroup of poxviruses. Arch. Virol. **54**, 143—145 (1977b).

43. BAXBY, D., GHABOOSI, B.: Laboratory characteristics of poxviruses isolated from captive elephants in Germany. J. gen. Virol. **37**, 407—414 (1977).

44. BAXBY, D., OSBORNE, A. D.: Antibody studies in natural bovine cowpox. J. Hyg. **83**, 425—428 (1979).
45. BAXBY, D., ASHTON, D. G., JONES, D., THOMSETT, L. R., DENHAM, E. M.: Cowpox virus infection in unusual hosts. Vet. Res. **109**, 175 (1979).
46. BAXBY, D., SHACKLETON, W. B., WHEELER, J., TURNER, A.: Comparison of cowpox-like viruses isolated from European zoos. Arch. Virol. **61**, 337—340 (1979).
47. BECKER, Y., JOKLIK, W. K.: Messenger RNA in cells infected with vaccinia virus. Biochem. **51**, 577—585 (1964).
48. BEDSON, H. S.: Multiple protein functions on the replication of poxvirus DNA. J. gen. Virol. **3**, 147—151 (1968).
49. BEN-HAMIDA, F., BEAUD, G.: *In vitro* inhibition of protein synthesis by purified cores from vaccinia virus. Proc. Natl. Acad. Sci. U.S.A. **75**, 175—179 (1978).
50. BERGOIN, M., DEVAUCHELLE, G., VAGO, C.: Electron microscopy of the pox-like virus of *Melolontha melolontha* L *(Coleoptera scarabeidae)*. Arch. ges. Virusforsch. **28**, 285—302 (1969).
51. BERGOIN, M., DEVAUCHELLE, G., VAGO, C.: Electron microscopy study of Melolontha poxvirus: the fine structure of occluded virions. Virol. **43**, 453—467 (1971).
52. BERGOIN, M., DALES, S.: Comparative observations on poxviruses of invertebrates and vertebrates. In: MARAMOROSCH, K., KURSTAK, E. (eds.), Comparative Virology, 171—203. New York-London: Academic Press 1971.
53. BERNS, K. I., SILVERMAN, C.: Natural occurrence of cross-linked vaccinia virus deoxyribonucleic acid. J. Virol. **5**, 299—304 (1970).
54. BERNS, K. I., SILVERMAN, C., WEISSBACH, A.: Separation of a new deoxyribonucleic acid polymerase from vaccinia-infected HeLa cells. J. Virol. **4**, 15—23 (1969).
55. BERGER, N. A., KAUFF, R. A., SIKORSKI, G. W.: ATP-independent DNA synthesis in vaccinia-infected L cells. Biochim. biophys. Acta **520**, 531—538 (1978).
56. BERRY, G. P., DEDRICK, H. M.: A method for changing the virus of rabbit-fibroma (Shope) into that of infectious myxomatosis *(Sanarelli)*. J. Bacteriol. **31**, 50—51 (1936).
57. BIALY, H. S., COLBY, C.: Inhibition of early vaccinia virus ribonucleic acid synthesis in interferon-treated chicken embryo fibroblasts. J. Virol. **9**, 286—289 (1972).
58. BILIMORIA, S. L., ARIF, B. M.: Subunit protein and alkaline protease of Entomopoxvirus spheroids. Virol. **96**, 596—603 (1979).
59. BLACKMANN K. E., BUBEL, H. C.: Origin of the vaccinia virus hemagglutinin. J. Virol. **9**, 290—296 (1972).
60. BLAND, J. O. W., ROBINOW, C. F.: The inclusion bodies of vaccinia and their relationship to the elementary bodies studied in cultures of the rabbit's cornea. J. Path. Bact. **48**, 381—403 (1939).
61. BODO, G., SCHEIRER, W., SUH, M., SCHULTZE, B., HORAK, I., JUNGWIRTH, C.: Protein synthesis in pox-infected cells treated with interferon. Virol. **50**, 140—147 (1972).
62. BOLDEN, A., PEDRALI-NOY, G., WEISSBACH, A.: Vaccinia virus infection of the HeLa cells. II. Disparity between cytoplasmic and nuclear viral-specific RNA. Virol. **94**, 138—145 (1979).
63. BOLLINGER, O.: Über *Epithelioma contagiosum* beim Haushuhn und die sogenannten Pocken des Geflügels. Arch. Path. Anat. Physiol. **58**, 349—388 (1873).
64. BOONE, R. F., MOSS, B.: Methylated 5′ terminal sequences of vaccinia virus mRNA species made *in vivo* at early and late times after infection. Virol. **79**, 67—80 (1977).
65. BOONE, R. F., ENSINGER, M. J., MOSS, B.: Synthesis of mRNA guanylyltransferase and mRNA methyltransferases in cells infected with vaccinia virus. J. Virol. **21**, 475—483 (1977).
66. BOONE, R. F., MOSS, B.: Sequence complexity and relative abundance of vaccinia virus mRNA's synthesized *in vivo* and *in vitro*. J. Virol. **26**, 554—569 (1978).

67. BOONE, R. F, PARR, R. P., MOSS, B.: Intermolecular duplexes formed from polyadenylated vaccinia virus RNA. J. Virol. **301**, 365—374 (1979).

68. BOSSART, W., PAOLETTI, E., NUSS, D. L.: Cell free translation of purified virion-associated molecular weight RNA synthesized *in vitro* by vaccinia virus. J. Virol. **28**, 905—916 (1978a).

69. BOSSART, W., NUSS, D. L., PAOLETTI, E.: Effect of UV irradiation on the expression of vaccinia virus gene products synthesized in a cell-free system coupling transcription and translation. J. Virol. **26**, 673—680 (1978b).

70. BOULTER, E. A., APPLEYARD, G.: Differences between extracellular and intracellular forms of poxvirus and their implications. Prog. med. Virol. **16**, 86—108 (1973).

71. BURGOYNE, R. D., STEPHEN, J.: Further studies on a vaccinia virus cytotoxin present in infected cell extracts: identification as surface tubule monomer and possible mode of action. Arch. Virol. **59**, 107—119 (1979).

72. CABRERA, C. V., ESTEBAN, M.: Procedure for purification of intact DNA from vaccinia virus. J. Virol. **25**, 442—445 (1978).

73. CABRERA, C. V., ESTEBAN, M., McCARRON, R., McALLISTER, W. T., HOLOWCZAK, J. A.: Vaccinia virus transcription: hybridization of mRNA to restriction fragments of vaccinia DNA. Virol. **86**, 102—114 (1978).

74. CAIRNS, J.: The initiation of vaccinia infection. Virol. **11**, 603—623 (1960).

75. CASSEL, W. A., FATER, B.: Immunizing properties of hemagglutinating vaccinia virus and non-hemagglutinating virus. Virol. **5**, 571—573 (1958).

76. CHALBERG, M. D., ENGLUND, P. T.: The DNA polymerase induced by vaccinia virus. J. biol. Chem. **254**, 7812—7819 (1979).

77. CHAN, J. C., HODES, M. E.: Mechanism of Shope fibroma virus induced suppression of host deoxyribonucleic acid synthesis. Infect. and Immun. **7**, 532—538 (1973).

78. CHANG, A., METZ, D. H.: Further investigations on the mode of entry of vaccinia virus into cells. J. gen. Virol. **32**, 275—282 (1976).

79. CHEEVERS, W. P., RANDALL, C. C.: Viral and cellular growth and sequential increase of protein and DNA during fowlpox infection *in vivo*. Proc. Soc. exp. Biol. Med. **127**, 401—405 (1968).

80. CHEEVERS, W. P., O'CALLAGHAN, D. J., RANDALL, C. C.: Biosynthesis of host and viral deoxyribonucleic acid during hyperplastic fowlpox infection *in vivo*. J. Virol. **2**, 421—429 (1968).

81. CHERNOS, V. I., BELANOV, E. G., VASILIEVA, N. N.: Temperature sensitive mutants of vaccinia virus. 1. Isolation and preliminary characterization. Acta Virol. **22**, 81—90 (1978).

82. CHO, C. T., WENNER, H. A.: Monkeypox virus. Bacteriol. Rev. **37**, 1—18 (1973).

83. CITARELLA, R. V., MULLER, R., SCHLABACH, A., WEISSBACH, A.: Studies on vaccinia virus-directed deoxyribonucleic acid polymerase. J. Virol. **10**, 721—729 (1972).

84. CLARK, E., NAGLER, F. P. O.: Haemagglutination by viruses. The range of susceptible cells with special reference to agglutination by vaccinia virus. Austr. J. exp. Biol. med. Sci. **21**, 103—106 (1943).

85. COHEN, G. H., WILCOX, W. C.: Soluble antigens of vaccinia infected mammalian cells. I. Separation of virus-induced soluble antigens into two classes on the basis of physical characteristics. J. Bact. **92**, 676—686 (1966).

86. COHEN, G. H., WILCOX, W. C.: Soluble antigens of vaccinia infected mammalian cells. III. Relation of "early" and "late" proteins to virus structure. J. Virol. **2**, 449—455 (1968).

87. COLBY, C., DUESBERG, P. H.: Double-stranded RNA in vaccinia virus infected cells. Nature **222**, 940—944 (1969).

88. COOKE, B. C., WILLIAMSON, J. D.: Enhanced utilization of citrulline in rabbitpox virus-infected mouse sarcoma 180 cells. J. gen. Virol. **21**, 339—348 (1973).

89. COOPER, R. J., BEDSON, H. S.: Temperature-sensitive events in the growth of alastrim virus in chick embryo cells. J. gen. Virol. **21**, 239—251 (1973).

90. COOPER, J. A., MOSS, B.: Transcription of vaccinia virus mRNA coupled to translation *in vitro*. Virol. **88**, 149—165 (1978).

91. COOPER, J. A., MOSS, B.: *In vitro* translation of immediate early and late classes of RNA from vaccinia virus-infected cells. Virol. **96**, 368—380 (1979).

92. CRAIGIE, J.: The nature of the vaccinia flocculation reaction and observations on the elementary bodies of vaccinia. Brit. J. exp. Path. **13**, 259—268 (1932).

93. CRAIGIE, J., WISHART, F. O.: Studies on the soluble precipitable substances of vaccinia. I. The dissociation *in vitro* of soluble precipitable substances from elementary bodies of vaccinia. J. exp. Med. **64**, 803—818 (1936a).

94. CRAIGIE, J., WISHART, F. O.: Studies on the soluble precipitable substances of vaccinia. II. The soluble precipitable substances of dermal vaccine. J. exp. Med. **64**, 819—830 (1936b).

95. CROUCH, N. A., HINZE, H. C.: Modification of cultured rabbit cells by ultraviolet-inactivated noncytocidal Shope fibroma virus (39843). Proc. Soc. exp. Biol. Med. **155**, 523—527 (1977).

96. DAHL, R., KATES, J. R.: Synthesis of vaccinia virus early and late mRNA *in vitro* with nucleoprotein structures isolated from infected cells. Virol. **42**, 463—472 (1970).

97. DALES, S.: An electron microscopic study of the early association between two mammalian viruses and their hosts. J. Cell Biol. **13**, 303—322 (1962).

98. DALES, S.: The uptake and development of vaccinia virus in strain L cells followed with labeled viral deoxyribonucleic acid. J. Cell Biol. **18**, 51—72 (1963).

99. DALES, S.: Relation between penetration of vaccinia, release of viral DNA, and initiation of genetic functions. In: POLLARD, M. (ed.), Perspectives in Virology IV, 47—71. New York: Harper & Row 1965a.

100. DALES, S.: Penetration of animal viruses into cells. In: MELNICK, J. L. (ed.), Prog. med. Virol. **7**, 1—43. New York-Basel: Karger 1965b.

101. DALES, S.: Effects of Streptovitacin A on the initial events in the replication of vaccinia and reovirus. Proc. Nat. Acad. Sci. U.S.A. **54**, 462—468 (1965c).

102. DALES, S.: Involvement of membranes in the infectious cycle of vaccinia. In: RICHTER, G. W., SCARPELLI, D. G. (ed.), Cell Membrane: Biological and Pathological Aspects, 136—144. Baltimore: Williams & Wilkins 1971.

103. DALES, S.: Early events in cell-animal virus interactions. Bact. Rev. **37**, 103—135 (1973).

104. DALES, S., SIMINOVITCH, L.: The development of vaccinia virus in Earle's L strain cells as examined by electron microscopy. J. biophys. biochem. Cytol. **10**, 475—503 (1961).

105. DALES, S., KAJIOKA, R.: The cycle of multiplication of vaccinia virus in Earle's strain L Cells. I. Uptake and penetration. Virol. **24**, 278—294 (1964).

106. DALES, S., MOSSBACH, E. H.: Vaccinia as a model for membrane biogenesis. Virol. **35**, 564—583 (1968).

107. DALES, S., STERN, W., WEINTRAUB, S. B., HUIMA, T.: Genetically controlled surface modifications by poxviruses influencing cell-cell and cell-virus interactions. In: BEERS, R. F., JR., BASSETT, E. G. (ed.), Miles 9th International Symposium Cell Membrane, 253—270. New York: Raven Press 1976.

108. DALES, S., MILOVANOVITCH, V., POGO, B. G. T., WEINTRAUB, S. B., HUIMA, T., WILTON, S., McFADDEN, G.: Biogenesis of vaccinia: isolation of conditional lethal mutants and electron microscopic characterization of their phenotypically expressed defects. Virol. **84**, 403—428 (1978).

109. DEFILIPPES, F. M.: Restriction enzyme digests of rapidly renaturing fragments of vaccinia virus DNA. J. Virol. **17**, 227—238 (1976).

110. DOHERTY, P. C., BENNINK, J. R.: Vaccinia-specific cytotoxic T-cell responses in the context of H-2 antigens not encountered in thymus may reflect aberrant recognition of a virus H-2 complex. J. exp. Med. **149**, 150—157 (1979).

111. DOWNER, D. N., ROGERS, H. W., RANDALL, C. C.: Endogenous protein kinase and phosphate acceptor proteins in vaccinia virus. Virol. **52**, 13—21 (1973).

112. DOWNIE, A. W.: A study of the lesions produced experimentally by cowpox virus. J. Path. Bact. **48**, 361—378 (1939).
113. DOWNIE, A. W., DUMBELL, K. R.: Pox viruses. Ann. Rev. Microbiol. **10**, 237—252 (1956).
114. DRILLIEN, R., TRIPIER, F., KOEHREN, F., KIRN, A.: A temperature-sensitive mutant of vaccinia virus defective in an early stage of morphogenesis. Virol. **79**, 369—380 (1977).
115. DRILLIEN, R., SPEHNER, D., KIRN, A.: Host range restriction of vaccinia virus in Chinese hamster ovary cells: relationship to shutoff of protein synthesis. J. Virol. **28**, 843—850 (1978).
116. DUBBS, D. R., KIT, S.: Isolation and properties of vaccinia mutants deficient in thymidine kinase-inducing activity. Virol. **22**, 214—225 (1964).
117. DUMBELL, K. R., DOWNIE, A. W., VALENTINE, R. C.: The ratio of the number of particles to infectious titer of cowpox and vaccinia virus suspensions. Virol. **4**, 467—482 (1957).
118. DURAN-REYNALS, M. L., STANLEY, B.: Vaccinia dermal infection and methy-cholanthrene in cortisone-treated mice. Science **134**, 1984—1985 (1961).
119. DUVAUCHELLE, G., BERGOIN, M., VAGO, C.: Etude ultrastructurale du cycle de réplication d'un entomopoxvirus dans les hémocytes de son hôte. Ultrast. Res. **37**, 301—321 (1971).
120. EASTERBROOK, K. B.: The multiplication of vaccinia virus in suspended K.B. cells. Virol. **15**, 404—416 (1961).
121. EASTERBROOK, K. B.: Interference with the maturation of vaccinia virus by Isatin β-thiosemicarbazone. Virol. **17**, 245—251 (1962).
122. EASTERBROOK, K. B.: Controlled degradation of vaccinia virions *in vitro*: an electron microscopic study. J. Ultrastruc. Res. **14**, 484—486 (1966).
123. ENSINGER, M. J., MARTEN, S. E., PAOLETTI, E., MOSS, B.: Modification of the 5′-terminus of mRNA by soluble guanylyl- and methyltransferases from vaccinia virus. Proc. Natl. Acad. Sci. U.S A **72**, 3385—3389 (1975)
124. EPSTEIN, M. A.: Purification using genetron extraction (fluorocarbon). Brit. J. exp. Pathol. **39**, 436—446 (1958a).
125. EPSTEIN, M. A.: Structural differentiation in the nucleoid of mature vaccinia virus. Nature **18**, 784—785 (1958b).
126. ERON, L. J., McAUSLAN, B. R.: The nature of poxvirus-induced deoxyribo-nucleases. Biochem. Biophys. Res. Comm. **22**, 518—523 (1966).
127. ESPOSITO, J. J., OBIJESKI, J. F., NAKANO, J. H.: Serological relatedness of monkeypox, variola, and vaccinia viruses. J. med. Virol. **1**, 35—47 (1977a).
128. ESPOSITO, J. J., OBIJESKI, J. F., NAKANO, J. H.: The virion and soluble antigen proteins of variola, monkeypox, and vaccinia viruses. J. med. Virol. **1**, 95—110 (1977b).
129. ESPOSITO, J. J., OBIJESKI, J. F., NAKANO, J. H.: Serological relatedness of monkeypox, variola, and vaccinia viruses. J. med. Virol. **1**, 35—47 (1977c).
130. ESPOSITO, J. J., OBIJESKI, J. F., NAKANO, J. H.: Orthopoxvirus DNA: Strain differentiation by electrophoresis of restriction endonuclease fragmented virion DNA. Virol. **89**, 53—66 (1978).
131. ESSANI, K., DALES, S.: Biogenesis of vaccinia: evidence for more than 100 poly-peptides in the virion. Virol. **95**, 385—394 (1979).
132. ESTEBAN, M., METZ, D. H.: Early virus protein synthesis in vaccinia virus-infected cells. J. gen. Virol. **19**, 201—216 (1973).
133. ESTEBAN, M., HOLOWCZAK, J. A.: Replication of vaccinia DNA in mouse L cells. I. *In vivo* DNA synthesis. Virol. **78**, 57—75 (1977a).
134. ESTEBAN, M., HOLOWCZAK, J. A.: Replication of vaccinia DNA in mouse L cells. II. *In vitro* DNA synthesis in cytoplasmic extracts. Virol. **78**, 76—86 (1977b).
135. ESTEBAN, M., HOLOWCZAK, J. A.: Replication of vaccinia DNA in mouse L cells. III. Intracellular forms of viral DNA. Virol. **82**, 308—322 (1977c).
136. ESTEBAN, M., FLORES, L., HOLOWCZAK, J. A.: Model for vaccinia virus DNA replication. Virol. **83**, 467—473 (1977).

137. ESTEBAN, M. HOLOWCZAK, J. A.: Replication of vaccinia in mouse L cells IV. Protein synthesis and viral replication. Virol. **86**, 376—390 (1978).
138. EWTON, D., HODES, M. E.: Nucleic acid synthesis in HeLa cells infected with Shope fibroma virus. Virol. **33**, 77—83 (1967).
139. FENGER, T., ROUHANDEH, H.: Proteins of Yaba monkey tumor virus. I. Structural proteins. J. Virol. **18**, 757—764 (1976).
140. FENNER, F.: Myxomatosis. Brit. med. Bull. **15**, 240—245 (1959).
141. FENNER, F.: White variants derived from poxviruses. World Health Organ. Workshop on smallpox, Atlanta, June 1979.
142. FENNER, F., SAMBROOK, J.: Conditional lethal mutants of rabbitpox virus II. Mutants (p) that fail to multiply in Pk-2a cells. Virol. **28**, 600—609 (1966).
142a. FENNER, F., WOODROOFE, G. M.: The reactivation of Poxviruses. II. The range of reactivating viruses. Virol. **11**, 185—201 (1960).
143. FENNER, F., McAUSLAN, B. R., MIMS, C. A., SAMBROOK, J., WHITE, D. O.: The biology of animal viruses, 2nd ed., 480—481. New York: Academic Press 1974.
144. FORTE, M. A., FANGMAN, W. L.: Naturally occurring cross-links in yeast chromosomal DNA. Cell **8**, 425—431 (1976).
145. GAFFORD, L. G., RANDALL, C. C.: The high molecular weight of the fowlpox virus genome. J. mol. Biol. **26**, 303—310 (1967).
145a. GAFFORD, L. G., RANDALL, C. C.: Further studies on high molecular weight fowlpox virus DNA and its hydrodynamic properties. Virol. **40**, 298—306 (1970).
146. GAFFORD, L. G., MITCHELL, E. B., RANDALL, C. C.: A comparison of sedimentation behavior of three poxvirus DNAs. Virol. **89**, 229—239 (1978).
147. GANGEMI, J. D., SHARP, D. G.: Host-induced influences on the syntheses of defective vaccinia particles. Virol. **73**, 165—172 (1976).
148. GARON, C. F., MOSS, B.: Glycoprotein synthesis in cells infected with vaccinia virus. II. A glycoprotein component of the virion. Virol. **46**, 233—246 (1971).
149. GARON, C. F., BARBOSA, E., MOSS, B.: Visualization of an inverted terminal repetition in vaccinia virus DNA. Proc. Natl. Acad. Sci. U.S.A. **75**, 4863—4867 (1978).
150. GAUSH, C. R., YOUNGNER, J. S.: Lipids of virus infected cells. II. Lipid analysis of HeLa cells infected with vaccinia virus. Proc. Soc. exp. Biol. Med. **112**, 1082—1085 (1963).
151. GAYLORD, W. H., MELNICK, J. L.: Intracellular forms of poxviruses as shown by the electron microscope. J. exp. Med. **98**, 157—172 (1954).
152. GEISTER, R. G., PETERS, D. H. A.: Quantitative use of the electron microscope in virus research. Method of analysing and predicting the biological titers of aggregated virus suspensions with a new law of aggregation. Lab. Invest. **14**, 864—874 (1969).
153. GERSHOWITZ, A., MOSS, B.: Abortive transcription products of vaccinia virus are guanylylated, methylated, and polyadenylylated. J. Virol. **31**, 849—853 (1979).
154. GESHELIN, P., BERNS, K. I.: Characterization and localization of the naturally occurring cross-links in vaccinia virus DNA. J. mol. Biol. **88**, 785—796 (1974).
155. GHENDON, Y. Z.: Conditional-lethal mutants of animal viruses. Progr. med. Virol. **14**, 68—122 (1972).
156. GHENDON, Y. Z., ROZINA, E. E., MARCHENKO, A.: On the 'neuropathogenicity' mechanism of vaccinia virus. Acta Virol. **8**, 359—368 (1973).
157. GINSBERG, A. H., JOHNSON, K. P.: Vaccinia virus meningitis in mice after intracerebral inoculation. Infect. Immun. **13**, 1221—1227 (1976).
158. GISPEN, R., BRAND-SAATHOF, B.: Three specific antigens produced in vaccinia, variola, and monkeypox infections. J. infect. Dis. **129**, 289—295 (1974).
159. GISPEN, R., BRAND-SAATHOF, B., HEKKER, A. C.: Monkeypox specific antibodies in human and simian sera from the Ivory Coast and Nigeria. Bull. World Health Organ. **53**, 355—360 (1976).
160. GOLD, P., DALES, S.: Localization of nucleotide phosphohydrolase activity within vaccinia. Proc. Nat. Acad. Sci. U.S.A. **60**, 845—852 (1968).

161. GRADY, L. J., PAOLETTI, E.: Molecular complexity of vaccinia DNA and the presence of reiterated sequences in the genome. Virol. **79**, 337—341 (1977).
162. GRANADOS, R. R.: Insect poxviruses: pathology, morphology, and development. Miscellaneous Publ. Entomol. Soc. Amer. **9**, 73—94 (1973a).
163. GRANADOS, R. R.: Entry of an insect poxvirus by fusion of the virus envelope with the host membrane. Virol. **52**, 305—309 (1973b).
164. GRANADOS, R. R., ROBERTS, D. W.: Electron microscopy of a poxlike virus infecting an invertebrate host. Virol. **40**, 230—243 (1970).
165. GREEN, R. H., ANDERSON, T. F., SMADEL, J. E.: Morphological structure of the virus of vaccinia. J. exp. Med. **75**, 651—655 (1942).
166. GREEN, M., PINA, M.: Stimulation of DNA synthesizing enzymes of cultured human cells by vaccinia virus infection. Virol. **17**, 603—604 (1962).
167. GRIMLEY, P. M., ROSENBLUM, E. N., MIMS, S. J., MOSS, B.: Interruption by rifampin of an early stage in vaccinia virus morphogenesis: accumulation of membranes which are precursors of virus envelopes. J. Virol. **6**, 519—533 (1970).
168. GUARNIERI, G.: Infezione vaccinia E variolosa. Archivio per le Scienze Med. **16**, 403—424 (1892).
169. HANAFUSA, T., HANAFUSA, H., KAMAHORA, J.: Transformation of ectromelia into vaccinia virus in tissue culture. Virol. **8**, 525—527 (1959).
170. HARFORD, G. G., HAMLIN, A., RIEDERS, E.: Electron microscopic autoradiography of DNA synthesis in cells infected with vaccinia virus. Exp. Cell Res. **42**, 50—57 (1966).
171. HARPER, J. M., PARSONAGE, M. T., PELHAM, H. R. B., DARBY, G.: Heat-inactivation of vaccinia virus particle-associated functions: properties of heated particles *in vivo* and *in vitro*. J. Virol. **26**, 646—659 (1978).
172. HARPER, L., BEDSON, H. S., BUCHAN, A.: Identification of Orthopoxviruses by polyacrylamide gel electrophoresis of intracellular polypeptides. Virol. **93**, 435—444 (1979).
173. HILLER, G., WEBER, K., SCHNEIDER, L., PARAJSZ, C., JUNGWIRTH, C.: Interaction of assembled progeny pox viruses with the cellular cytoskeleton. Virol. **98**, 142—153 (1979).
174. HIMMELWEIT, F.: Observations on living vaccinia and ectromelia viruses by high power microscopy. Brit. J. exp. Path. **19**, 108—123 (1938).
175. HIRST, G. K.: Adsorption of influenza virus on cells of the respiratory tract. J. exp. Med. **19**, 177—184 (1943).
176. HOAGLAND, C. L., LAVIN, G. I., SMADEL, J. E., RIVERS, T. M.: Constituents of elementary bodies of vaccinia. II. Properties of nucleic acid obtained from vaccine virus. J. exp. Med. **72**, 139—147 (1940).
177. HODGSON, J., WILLIAMSON, J. D.: Ornithine decarboxylase activity in uninfected and vaccinia virus-infected HeLa cells. Biochem. Biophys. Res. Comm. **63**, 308—312 (1975).
178. HOLOWCZAK, J. H.: Glycopeptides of vaccinia virus. I. Preliminary characterization and hexosamine content. Virol. **42**, 87—99 (1970).
179. HOLOWCZAK, J. H.: Uncoating of poxviruses. I. Detection and characterization of subviral particles in the uncoating process. Virol. **50**, 216—232 (1972).
180. HOLOWCZAK, J.: Poxvirus DNA. I. Studies on the structure of the vaccinia genome. Virol. **72**, 121—133 (1976).
181. HOLOWCZAK, J. A., JOKLIK, W. K.: Studies on the structural proteins of vaccinia virus. I. Structural proteins of virions and cores. Virol. **33**, 717—725 (1967).
182. HOLOWCZAK, J. A., DIAMOND, L.: Poxvirus DNA. II. Replication of vaccinia virus DNA in the cytoplasm of the HeLa cells. Virol. **72**, 134—146 (1976).
183. HRUBY, D. E., LYNN, D. L., KATES, J. R.: Vaccinia virus replication requires active participation of the host cell transcriptional apparatus. Proc. Natl. Acad. Sci. U.S.A. **76**, 1887—1890 (1979).
184. HSU, H. T., MCBEATH, J. H., BLACK, L. M.: The comparative susceptibilities of cultured vector and nonvector leafhopper cells to three plant viruses. Virol. **81**, 257—262 (1977).

185. HYDE, J. M., GAFFORD, L. G., RANDALL, C. C.: Molecular weight determination of fowlpox virus DNA by electron microscopy. Virol. **33**, 112—129 (1967).
186. ICHIHASHI, Y., MATSUMOTO, S.: The relationship between poxvirus and A-type inclusion body during double infection. Virol. **36**, 262—270 (1968a).
187. ICHIHASHI, Y., MATSUMOTO, S.: Genetic character of poxvirus A-type inclusion. Virus (Kyoto) **18**, 237—247 (1968b).
188. ICHIHASHI, Y., DALES, S.: Biogenesis of poxviruses: interrelationship between hemagglutinin production and polykaryocytosis. Virol. **46**, 533—543 (1971).
189. ICHIHASHI, Y., MATSUMOTO, S., DALES, S.: Biogenesis of poxviruses: Role of A-type inclusions and host cell membranes in virus dissemination. Virol. **46**, 507—532 (1971).
190. ICHIHASHI, Y., DALES, S.: Biogenesis of poxviruses: relationship between a translation complex and formation of A-type inclusions. Virol. **51**, 297—319 (1973).
191. IKUTA, K., MIYAMOTO, H., KATO, S.: Studies on the polypeptides of poxvirus. I. Comparison of structural polypeptides in vaccinia, cowpox and Shope fibroma viruses. Biken J. **21**, 51—61 (1978a).
192. IKUTA, K., MIYAMOTO, H., KATO, S.: Studies on the polypeptides of poxvirus. II. Comparison of virus-induced polypeptides in cells infected with vaccinia, cowpox and Shope fibroma viruses. Biken J. **21**, 77—94 (1978b).
193. IKUTA, K., MIYAMOTO, H., KATO, S.: Two components specifically responsible for hemagglutination in vaccinia and cowpox viruses. Virol. **96**, 327—331 (1979).
194. ITO, M. BARRON, A. L.: Studies on a strain of vaccinia virus defective in surface antigen production (36461). Proc. Soc. exp. Biol. Med. **140**, 374—377 (1972).
195. JACQUEMONT, B., GRANGE, J., GAZZOLO, L., RICHARD, M. H.: Composition and size of Shope fibroma virus deoxyribonucleic acid. J. Virol. **9**, 836—841 (1972).
196. JAUREGUIBERRY, G.: Cleavage of vaccinia virus DNA by restriction endonuclease Bal I, EcoRI, BamHI. Isolation of the natural cross-links. FEBS lett. **83**, 111—117 (1977).
197. JAUREGUIBERRY, G., BEN-HAMIDA, F., CHAPEVILLE, F., BEAUD, G.: Messenger activity of RNA transcribed *in vitro* by DNA-RNA polymerase associated to vaccinia virus cores. J. Virol. **15**, 1467—1474 (1975).
198. JOKLIK, W. K.: The multiplication of poxvirus DNA. Cold Spring Harb. Symp. quant. Biol. **27**, 199—208 (1962a).
199. JOKLIK, W. K.: Reactivation of poxviruses: fate of reactivable virus within the cell. Nature **196**, 556—558 (1962b).
200. JOKLIK, W. K.: The preparation and characteristics of highly purified radioactively labelled poxvirus. Biochim. biophys. Acta **61**, 290—301 (1962c).
201. JOKLIK, W. K.: Some properties of poxvirus deoxyribonucleic acid. J. mol. Biol. **5**, 265—274 (1962d).
202. JOKLIK, W. K.: The intracellular uncoating of poxvirus DNA. I. The fate of radioactively-labeled rabbitpox virus. J. mol. Biol. **8**, 263—276 (1964).
203. JOKLIK, W. K.: The poxviruses. Bact. Rev. **30**, 33—66 (1966).
204. JOKLIK, W. K., WOODROOFE, G. M., HOLMES, I. H., FENNER, F.: The reactivation of poxviruses: I. Demonstration of the phenomenon and techniques of assay. Virol. **11**, 168—184 (1960a).
205. JOKLIK, W. K., HOLMES, I. H., BRIGGS, M. J.: The reactivation of poxviruses. III. Properties of reactivable particles. Virol. **11**, 202—218 (1960b).
206. JOKLIK, W. K., BECKER, Y.: The replication and coating of vaccinia DNA. J. mol. Biol. **10**, 452—474 (1964).
207. JOKLIK, W. K., BECKER, Y.: Studies on the genesis of ribosomes. II. The association of nascent mRNA with the 40S subribosomal particle. J. mol. Biol. **13**, 511—520 (1965).
208. JOKLIK, W. K., MERRIGAN, T. C.: Concerning the mechanism of action of interferon. Proc. Nat. Acad. Sci. U.S.A. **56**, 558—564 (1966).
209. JUNGWIRTH, C., JOKLIK, W. K.: Studies on early enzymes in HeLa Cells infected with vaccinia virus. Virol. **27**, 80—93 (1965).

210. KAERLEIN, M., HORAK, I.: Phosphorylation of ribosomal proteins in HeLa cells infected with vaccinia virus. Nature **259**, 150—151 (1976).

211. KAERLEIN, M., HORAK, I.: Identification and characterization of ribosomal proteins phosphorylated in vaccinia virus infected HeLa cells. Eur. J. Biochem. **90**, 463—469 (1978).

212. KAJIOKA, R., SIMINOVITCH, L., DALES, S.: The cycle of multiplication of vaccinia virus in Earle's strain L cells. II. Initiation of DNA synthesis and morphogenesis. Virol. **24**, 295—309 (1964).

213. KAKU, H., KAMAHORA, J.: Giant cell formation in L cells infected with active vaccinia virus. Biken J. **6**, 299—315 (1964).

214. KATES, J.: Transcription of the vaccinia virus genome and the occurrence of polyriboadenylic acid sequences in Messenger RNA. Cold Spring Harb. Symp. quant. Biol. **35**, 743—751 (1970).

215. KATES, J., BEESON, J.: Ribonucleic acid synthesis in vaccinia virus. I. The mechanism of synthesis and release of RNA in vaccinia cores. J. mol. Biol. **50**, 1—18 (1970a).

216. KATES, J., BEESON, J.: Ribonucleic acid synthesis in vaccinia virus. II. Synthesis of polyriboadenylic acid. J. mol. Biol. **50**, 19—23 (1970b).

217. KATES, J. R., McAUSLAN, B.: Messenger RNA synthesis by a 'coated' viral genome. Proc. Natl. Acad. Sci. U.S.A. **57**, 314—320 (1967a).

218. KATES, J. R., McAUSLAN, B. R.: Relationship between protein synthesis and viral DNA synthesis. J. Virol. **1**, 110—114 (1967b).

219. KATES, J. R., McAUSLAN, B. R.: Poxvirus DNA-dependent RNA polymerase. Proc. Nat. Acad. Sci. U.S.A. **58**, 134—141 (1967c).

220. KATES, J., DAHL, R., MIELKE, M.: Synthesis and intracellular localization of vaccinia virus deoxyribonucleic acid-dependent RNA polymerase. J. Virol. **2**, 894—900 (1968).

221. KATO, S., AOYAMA, Y., KAMAHORA, J.: Phase contrast microscopy and phase contrast cinematography of viral inclusions produced by poxvirus. Biken J. **5**, 253—258 (1962a).

222. KATO, S., KURISU, M., KAMAHORA, J.: Effects of cytoplasmic DNA synthesis upon nuclear DNA synthesis in poxvirus-infected cells. Biken J. **5**, 227—231 (1962b).

223. KATO, S., KAMAHORA, J.: The significance of the inclusion formation of poxvirus group and herpes simplex virus. Symp. Cell Chem. **12**, 47—90 (1962).

224. KATO, S., HARA, J., OGAWA, M., MIYAMOTO, H., KAMAHORA, J.: Inclusion markers of cowpox viruses and Alastrim virus. Biken J. **6**, 233—235 (1963a).

225. KATO, S., AOYAMA, Y., KAMAHORA, J.: Autoradiography of the tissues of mice infected with ectromelia virus using ^{3}H-thymidine. Biken J. **6**, 9—16 (1963b).

226. KATZ, E., MOSS, B.: Synthesis of vaccinia viral proteins in cytoplasmic extracts. I. Incorporation of radioactively labeled aminoacids into polypeptides. J. Virol. **4**, 416—422 (1969).

227. KATZ, E., MOSS, B.: Formation of a vaccinia virus structural polypeptide from a higher molecular weight precursor: inhibition by rifampicin. Proc. Nat. Acad. Sci. U.S.A. **66**, 677—684 (1970a).

228. KATZ, E., MOSS, B.: Vaccinia virus structural polypeptide derived from a high molecular weight precursor: formation and integration into virus particles. J. Virol. **6**, 717—726 (1970b).

229. KATZ, E., MARGALITH, E., WINER, B.: An isatin β-thiosemicarbazone (IBT)-dependent mutant of vaccinia virus: the nature of the IBT-dependent step. J. gen. Virol. **21**, 477—482 (1973a).

230. KATZ, E., MARGALITH, E., WINER, B., LAZAR, A.: Characterization and mixed infections of three strains of vaccinia virus: wild type, IBT-resistant and IBT-dependent mutants. J. gen. Virol. **21**, 469—475 (1973b).

231. KATZ, E., MARGALITH, E., WINER, B.: Synthesis of vaccinia virus polypeptides in the presence of hydroxyurea. Antimicrob. Agents Chemother. **6**, 647—650 (1974).

232. KAVERIN, N. V., VARICH, N. L., SURGAY, V. V., CHERNOS, V. I.: A quantitative estimation of poxvirus genome fraction transcribed as early and late mRNA. Virol. **65**, 112—119 (1975).
233. KELLER, F., DRILLIEN, R., KIRN, A.: Effect of cell-mediated immune factors on the replication of an attenuated temperature-sensitive mutant of vaccinia virus. Infect. Immun. **26**, 841—847 (1979).
234. KERR, I. M., FRIEDMAN, R. M., BROWN, R. E., BALL, L. A., BROWN, J. C.: Inhibition of protein synthesis in cell-free systems from interferon-treated infected cells: further characterization and effect of formylmethionyl tRNA. J. Virol. **13**, 9—21 (1974).
235. KILHAM, L., LERNER, E., HIATT, C., SHACK, J.: Properties of myxoma transforming agent. Proc. Soc. exp. Biol. Med. **98**, 689—692 (1958).
236. KIT, S., DUBBS, D. R.: Biochemistry of vaccinia-infected mouse fibroblasts (strain L—M). I. Effects on nucleic acid and protein synthesis. Virol. **18**, 274—285 (1962a).
237. KIT, S., DUBBS, D. R.: Biochemistry of vaccinia-infected mouse fibroblasts (strain L—M). II. Properties of the chromosomal DNA of infected cells. Virol. **18**, 286—293 (1962b).
238. KIT, S., DUBBS, D. R.: Biochemistry of vaccinia-infected mouse fibroblasts (strain L—M). IV. ^{3}H-thymidine uptake into DNA of cells exposed to cold shock. Exp. Cell Res. **31**, 397—406 (1963).
239. KIT, S., DUBBS, D. R., HSU, T. C.: Host cell DNA synthesis suppressed by vaccinia infection. Virol. **19**, 13—22 (1963a).
240. KIT, S., DUBBS, D. R., PIEKARSKI, L. J., HSU, T. C.: Deletion of thymidine kinase activity from L cells resistant to bromodeoxyuridine. Exp. Cell Res. **31**, 297—312 (1963b).
241. KIT, S., PIEKARSKI, L. J., DUBBS, D. R.: Induction of thymidine kinase by vaccinia-infected mouse fibroblasts. J. mol. Biol. **6**, 22—33 (1963c).
242. KIT, S., PIERKARSKI, L. J., DUBBS, D. R.: Effects of 5-fluorouracil, actinomycin D and mitomycin C on the induction of thymidine kinase by vaccinia-infected L-cells. J. mol. Biol. **7**, 497—510 (1963d).
243. KIT, S., VALLADARES, Y., DUBBS, D. R.: Effects of age of culture and vaccinia infection on uridine kinase activity of L-cells. Exp. Cell Res. **34**, 257—265 (1964).
244. KIT, S., JORGENSEN, G. N., LIAV, A., ZASLAVSKY, V.: Purification of vaccinia virus-induced thymidine kinase activity from ^{35}S-methionine-labeled cells. Virol. **77**, 661—676 (1977).
245. KITAMURA, T., OGATA, M.: Sensitivity of *Mastomys* and rabbit to the intranasal inoculation of monkeypox virus — their possible use as the sentinel animals in the ecological study of the monkeypox virus. W.H.O. Workshop on Poxviruses, Atlanta, Georgia, 1979.
246. KLEIMAN, J., MOSS, B.: Protein kinase activity from vaccinia virions: solubilization and separation into heat-labile and heat-stable components. J. Virol. **12**, 684—689 (1973).
247. KOSZINOWSKI, U., KRUSE, F., THOMSSEN, R.: Interactions between vaccinia virus and sensitized macrophages *in vitro*. Arch. Virol. **48**, 335—345 (1975).
248. LACOLLA, P., WEISSBACH, A.: Vaccinia virus infection of HeLa cells. I. Synthesis of vaccinia DNA in host cell nuclei. J. Virol. **15**, 305—315 (1975).
249. LAEMMLI, U. K.: Cleavage of structural proteins during the assembly of the head of bacteriophage T4. Nature **227**, 680—684 (1970).
250. LAKE, J. R., SILVER, M., DALES, S.: Biogenesis of vaccinia: complementation and recombination analysis of one group of conditional-lethal mutants defective in envelope self-assembly. Virol. **96**, 9—20 (1979).
251. LAMBERT, D., MAGEE, W. E.: Characterization of vaccinia virus DNA *in vitro*. Virol. **79**, 342—354 (1977).
252. LANGER, W. L.: Immunization against smallpox before JENNER. Scientific American **234**, 112—117 (1976).

253. LANZER, W., HOLOWCZAK, J. A.: Polyamines in vaccinia virions and polypeptides released from viral cores by acid extraction. J. Virol. **16**, 1254—1264 (1975).

254. LAPLACA, M.: On the mechanism of the cytopathic changes produced in human amnion cell cultures by the molluscum contagiosum virus. Arch. ges. Virusforsch. **18**, 374—378 (1966).

255. LAPLACA, M., PORTOLANI, M., MANNINI-PALENZONA, A., BARBANTI-BRODANO, G., BERNARDINI, A.: Further studies on the mechanism of the cytopathic changes produced by the molluscum contagiosum virus into human amnion cell cultures. Giorn. Microbiol. **15**, 205—216 (1967).

256. LEDINGHAM, J. C. G., ABERD, M. B.: The aetiological importance of the elementary bodies in vaccinia and fowlpox. Lancet **221**, 525—526 (1931).

257. LEVINE, S., MAGEE, W. E., HAMILTON, R. D., MILLER, O. V.: Effect of interferon on early enzyme and viral DNA synthesis in vaccinia virus infection. Virol. **32**, 33—40 (1967).

258. LOH, P., RIGGS, J. L.: Demonstration of the sequential development of vaccinial antigens and virus in infected cells: observations with cytochemical and differential fluorescent procedures. J. exp. Med. **114**, 149—160 (1961).

259. LYLES, D. S., RANDALL, C. C., GAFFORD, L. G., WHITE, H. B., JR.: Cellular fatty acids during fowlpox virus infection of three different host systems. Virol. **70**, 227—229 (1976).

260. MACKETT, M., ARCHARD, L. C.: Conservation and variation in Orthopoxvirus genome structure. J. gen. Virol. **45**, 683—701 (1979).

261. MAGEE, W. E.: DNA polymerase and deoxyribonucleotide kinase activities in cells infected with vaccinia virus. Virol. **17**, 604—607 (1962).

262. MAGEE, W. E., SAGIK, B. P.: The synthesis of deoxyribonucleic acid by HeLa cells infected with vaccinia virus. Virol. **8**, 134—137 (1959).

263. MAGEE, W. E., SHEEK, M. R., BURROUS, M. J.: The synthesis of vaccinial deoxyribonucleic acid. Virol. **11**, 296—299 (1960).

264. MAGEE, W. E., MILLER, O. V.: Dissociation of the synthesis of host and viral deoxyribonucleic acid. Biochim. biophys. Acta **55**, 818—826 (1962).

265. MAGEE, W. E., MILLER, O. V.: Immunological evidence for the appearance of a new DNA-polymerase in cells infected with vaccinia virus. Virol. **31**, 64—69 (1967).

266. MAGEE, W. E., MILLER, O. V.: Initiation of vaccinia virus infection in actinomycin D-pretreated cells. J. Virol. **2**, 678—685 (1968).

267. MAGEE, W. E., LEVINE, S., MILLER, O. V., HAMILTON, R. D.: Inhibition by interferon of the uncoating of vaccinia virus. Virol. **35**, 505—511 (1968).

268. MAHNEL, H.: Labordifferenzierung der Orthopockenviren. Zbl. vet. Med. **B21**, 242—258 (1974).

269. MARCHAL, J.: Infectious ectromelia. A hitherto undescribed virus disease in mice. J. Path. Bact. **33**, 713—728 (1930).

270. MARENNIKOVA, S. S., SHELUKHINA, E. M.: Susceptibility of some rodent species to monkeypox virus, and course of the infection. Bull. World Health Organ. **53**, 13—20 (1976).

271. MARENNIKOVA, S. S., MALTSEVA, N. N., KORNEEVA, V. I., GARANINA, M.: Outbreak of pox disease among Carnivora (Felidae) and Edentata. J. infect. Dis. **135**, 358—366 (1977).

272. MARENNIKOVA, S. S., SHELUKHINA, E. M.: Whitepox virus isolated from hamsters inoculated with monkeypox virus. Nature **276**, 291—292 (1978).

273. MARENNIKOVA, S. S., SHELUKHINA, E. M., MALTSEVA, N. N.: Monkeypox virus and whitepox viruses. Acta Virol. **22**, 512 (1978).

274. MARENNIKOVA, S. S., SHELUKHINA, E. M., MALTSEVA, N. N., MATSEVICH, G. R.: Monkeypox virus as a source of whitepox viruses. Intervirol. **11**, 333—340 (1979).

275. MARTIN, S. A., PAOLETTI, E., MOSS, B.: Purification of mRNA guanylyltransferase and mRNA (guanine-7-)-methyltransferase from vaccinia virions. J. biol. Chem. **250**, 9322—9329 (1975).

276. MARTIN, S. A., MOSS, B.: Modification of mRNA by mRNA guanylyltransferase and mRNA (guanine-7-)-methyltransferase from vaccinia virions. J. biol. Chem. **250**, 9330—9335 (1975).

277. MATSUMOTO, S.: On the proliferation of ectromelia virus in Ehrlich ascites tumor cells. An electron microscopic and cytochemical study. Acta Sch. Med. Univ. Kyoto **34**, 41—64 (1956).

278. MATTHEWS, R. E. F.: Classification and nomenclature of viruses. Intervirol. **12**, 160—164 (1979).

279. MAYR, A., MAHNEL, H.: Charakterisierung eines vom Rhinozeros isolierten Hühnerpockenvirus. Arch. ges. Virusforsch. **31**, 51—60 (1970).

280. MAYR, A., MAHNEL, H., MUNZ, E.: Systematisierung und Differenzierung der Pockenviren. Zbl. vet. Med. **B19**, 69—88 (1972).

281. MAYR, A., STICKL, H., MULLER, H. K., DANNER, K., SINGER, H.: Der Pockenimpfstamm MVA: Marker, genetische Struktur, Erfahrungen mit der parenteralen Schutzimpfung und Verhalten im abwehrgeschwächten Organismus. Zbl. Bakt. Hyg., I. Abt. Orig. **B167**, 375—390 (1978).

282. MBUY, G., BUBEL, H. C.: Concanavalin A-mediated cell agglutinability induced by vaccinia virions. Virol. **91**, 256—266 (1978).

283. MBUY, G. N., BUBEL, H. C., MORRIS, R. E.: Inhibition of cellular protein synthesis by vaccinia virus surface tubules. (Personal communication.)

284. MEDZON, E. L., BAUER, H.: Structural features of vaccinia virus revealed by negative staining, sectioning, and freeze-etching. Virol. **40**, 860—867 (1970).

285. MENNA, A., WYLER, R.: Comparison of five poxvirus genomes by analysis with restriction endonucleases HindIII, BamI, EcoRI. J. gen. Virol. **38**, 161—166 (1977).

286. MENNA, A., WITTEK, R., BACHMAN, P. A., MAYR, A., WYLER, R.: Physical characterization of 2 stomatitis papulosa virus genomes: A cleavage map for the restriction endonucleases HindIII and EcoRI. Arch. Virol. **59**, 145—156 (1979).

287. METZ, D. H., ESTEBAN, M., DANIELESCU, G.: The formation of virus polyribosomes in L cells infected with vaccinia virus. J. gen. Virol. **27**, 181—195 (1975a).

288. METZ, D. H., ESTEBAN, M., DANIELESCU, G.: The effect of interferon on the formation of virus polyribosomes in L cells infected with vaccinia virus. J. gen. Virol. **27**, 197—209 (1975b).

289. MILLER, G., ENDERS, J. F.: Vaccinia virus replication and cytopathic effect in cultures of phytohemagglutinin-treated human peripheral blood leukocytes. J. Virol. **2**, 787—792 (1968).

290. MILNER, R. J., BEATON, C. D.: An entomopoxvirus from *Oncopera alboguttata* (*Lepidoptera: Hepialidae*) in Australia. Intervirol. **11**, 341—350 (1979).

291. MILO, G. E., JR., YOHN, D. A.: Alterations of enzymes associated with plasma membranes and cellular organelles during infection of CV-1 cells with Yaba tumor poxvirus. Cancer Res. **35**, 199—206 (1975).

292. MIMS, C. A.: The response of mice to large intravenous injections of ectromelia virus. I. The fate of injected virus. J. exp. Pathol. **40**, 533—542 (1959).

293. MIMS, C. A.: Intracerebral injections and the growth of viruses in the mouse brain. Brit. J. exp. Pathol. **41**, 52—59 (1960).

294. MIMS, C. A.: Aspects of the pathogenesis of virus diseases. Bact. Rev. **28**, 30—71 (1964).

295. MONROY, G., SPENCER, E., HURWITZ, J.: Purification of mRNA guanylyltransferase from vaccinia virions. J. biol. Chem. **253**, 4481—4489 (1978a).

296. MONROY, G., SPENCER, E., HURWITZ, J.: Characteristics of reactions catalyzed by purified guanylyltransferase from vaccinia virus. J. biol. Chem. **253**, 4490 to 4498 (1978b).

297. MORGAN, C.: Vaccinia virus reexamined: development and release. Virol. **73**, 43—58 (1976a).

298. MORGAN, C.: The insertion of DNA into vaccinia virus. Science **193**, 591—592 (1976b).

299. MORGAN, C., ELLISON, S. A., ROSE, R. M., MOORE, D. H.: Structure and development of viruses observed in the electron microscope. II. Vaccinia and fowlpox viruses. J. exp. Med. **100**, 301—309 (1954).
300. MORITA, M., AOYAMA, Y., ARITA, M., AMONO, H., YOSHIZAWA, H., HASHIZUME, S., KOMATSU, T., TAGAYA, I.: Comparative studies of several vaccinia virus strains by intrathalamic inoculation into Cynomolgus monkeys. Arch. Virol. **53**, 197—208 (1977).
301. MOSS, B.: Inhibition of HeLa cell protein synthesis by the vaccinia virion. J. Virol. **2**, 1028—1037 (1968).
302. MOSS, B.: Reproduction of poxviruses. In: FRAENKEL-CONRAT, H., WAGNER, R. R. (ed.), Comprehensive Virology, Vol. 3, 405—474. New York: Plenum Press 1974.
303. MOSS, B., SALZMAN, N. P.: Sequential protein synthesis following vaccinia virus infection. J. Virol. **2**, 1016—1027 (1968).
304. MOSS, B., KATZ, E.: Synthesis of vaccinia viral proteins in cytoplasmatic extracts. II. Identification of early and late viral proteins. J. Virol. **4**, 596—602 (1969).
305. MOSS, B., FILLER, R.: Irreversible effects of cycloheximide during the early period of vaccinia virus replication. J. Virol. **5**, 99—108 (1970).
306. MOSS, B., ROSENBLUM, E. N., GARON, C. F.: Glycoprotein synthesis in cells infected with vaccinia virus. Virol. **46**, 221—232 (1971).
307. MOSS, B., ROSENBLUM, E. N.: Protein cleavage and poxvirus morphogenesis: tryptic peptide analysis of core precursors accumulated by blocking assembly with rifampicin. J. mol. Biol. **81**, 267—269 (1973).
308. MOSS, B., ROSENBLUM, E. N., PAOLETTI, E.: Polyadenylate polymerase from vaccinia virions. Nature new Biol. **245**, 59—63 (1973).
309. MOSS, B., ROSENBLUM, E. N.: Vaccinia virus polyriboadenylate polymerase: covalent linkage of the product with polyribonucleotide and polydeoxyribonucleotide. primers. J. Virol. **14**, 86—98 (1974).
310. MOSS, B., ROSENBLUM, E. N., GERSHOWITZ, A.: Characterization of a polyriboadenylate polymerase from vaccinia virus. J. biol. Chem. **250**, 4722—4729 (1975).
311. MOSS, B., GERSHOWITZ, A., WEI, C. M., BOONE, R. F.: Formation of guanylylated and methylated 5'-terminus of vaccinia virus mRNA. Virol. **72**, 341—351 (1976).
312. MOYER, R. W., ROTHE, C. T.: The white pock mutants of rabbit poxvirus. I. Spontaneous host range mutants contain deletions. Virol. **102**, 119—132 (1980).
313. MULLER, G.: Disaggregation of vaccinia virus with ultrasonic cleaners. Arch. Virol. **51**, 365—367 (1976).
314. MÜLLER, H. K., WITTEK, R., SCHAFFNER, W., SCHÜMPERLI, D., MENNA, A., WYLER, R.: Comparison of five poxvirus genomes by analysis with restriction endonucleases HindIII, BamI and EcoRI. J. gen. Virol. **38**, 135—147 (1977).
315. MUNYON, W. H., KIT, S.: Induction of cytoplasmic ribonucleic acid (RNA) synthesis in vaccinia-infected LM cells during inhibition of protein synthesis. Virol. **29**, 303—309 (1966).
316. MUNYON, W. E., PAOLETTI, E., GRACE, J. T., JR.: RNA polymerase activity in purified vaccinia virus. Proc. Nat. Acad. Sci. U.S.A. **58**, 2280—2287 (1967).
317. MUNYON, W., PAOLETTI, E., OSPINA, J., GRACE, J. T., JR.: Nucleotide phosphohydrolase in purified vaccinia virus. J. Virol. **2**, 167—172 (1968).
318. MUNYON, W., MANN, J., GRACE, T. J., JR.: Protection of vaccinia from heat inactivation by nucleotide triphosphates. J. Virol. **5**, 32—38 (1970).
319. MCAUSLAN, B. R.: The induction and repression of thymidine kinase in the poxvirus-infected HeLa Cell. Virol. **21**, 383—389 (1963a).
320. MCAUSLAN, B. R.: Control of induced thymidine kinase activity in the poxvirus infected cell. Virol. **20**, 162—168 (1963b).
321. MCAUSLAN, B. R.: Deoxyribonuclease activity of normal and poxvirus-infected HeLa cells. Biochim. Biophys. Res. Comm. **19**, 15—20 (1965).
322. MCAUSLAN, B. R., JOKLIK, W. K.: Stimulation of the thymidine phosphorylating system in HeLa cells on infection with poxvirus. Biochem. Biophys. Res. Comm. **8**, 486—491 (1962).

323. McAuslan, B. R., Kates, J. R.: Poxvirus-induced acid deoxyribonuclease: regulation of synthesis; control of activity *in vivo*; purification and properties of the enzyme. Virol. **33**, 709—716 (1967).

324. McCarron, R. J., Cabrera, C. V., Esteban, M., McAllister, W. T., Holowczak, J. A.: Structure of vaccinia DNA: analysis of the viral genome by restriction endonucleases. Virol. **86**, 88—101 (1978).

325. McCarthy, W. J., Granados, R. R., Roberts, D. W.: Isolation and characterization of Entomopox virions from virus containing inclusions of *Amsacta moorei (Lepidoptera: Arctudae)*. Virol. **59**, 59—69 (1974).

326. McClain, M. E.: The host range and plaque morphology of rabbitpox virus (RPut) and its μ mutants on chick fibroblasts, Pk-2a, and L9-9 cells. Austr. J. exp. Biol. med. Sci. **43**, 31—44 (1965).

327. McCrae, M. A., Pennington, T. H.: Specific secretion of polypeptides from cells infected with vaccinia virus. J. Virol. **28**, 828—834 (1978).

328. McFadden, G., Pace, W. E., Purres, J., Dales, S.: Biogenesis of poxviruses: transitory expression of *Molluscum contagiosum* early functions. Virol. **94**, 297—313 (1979).

329. McFadden, G., Dales, S.: Biogenesis of poxviruses: mirror image deletions in vaccinia virus DNA. Cell **18**, 101—108 (1979).

330. McFadden, G., Dales, S.: Biogenesis of poxviruses: preliminary characterization of conditional lethal mutants of vaccinia virus defective in DNA synthesis. Virol. **103**, 68—79 (1980).

330a. McFadden, G., Essani, K., Dales, S.: A new endonuclease restriction site which is at the locus of a temperature-sensitive mutation in vaccinia virus is associated with true and pseudoreversion. Virol. **101**, 277—280 (1980).

331. McKay Brown, Dorson, J. W., Bollum, F. J.: Terminal riboadenylate transferase: a polyA polymerase in purified vaccinia virus. J. Virol. **12**, 203—208 (1973).

331a. McRae, M. A., Szilagyi, J. F.: Preparation and characterization of a subviral particle of vaccinia virus containing DNA-dependent RNA polymerase activity. Virol. **68**, 234—244 (1975).

332. Nagayama, A., Pogo, B. G. T., Dales, S.: Biogenesis of vaccinia; separation of early stages from maturation by means of rifampicin. Virol. **40**, 1039—1051 (1970).

333. Nagington, J., Newton, A. A., Horne, R. W.: The structure of orf virus. Virol. **23**, 461—472 (1964).

334. Nevins, J. R., Joklik, W. K.: Poly(A) sequences of vaccinia virus mRNA: nature, mode of action and function during translation *in vitro* and *in vivo*. Virol. **63**, 1—14 (1975).

335. Nevins, J. R., Joklik, W. K.: Isolation and properties of the vaccinia virus-DNA-dependent RNA polymerase. J. biol. Chem. **252**, 6930—6938 (1977a).

336. Nevins, J. R., Joklik, W. K.: Isolation and partial characterization of the poly(A) polymerases from HeLa cells infected with vaccinia virus. J. biol. Chem. **252**, 6939—6947 (1977b).

337. Nishimura, C.: Properties of nucleic acids in a purified vaccinia virus and in vaccinia-infected HeLa cells. Japan J. med. Sci. Biol. **18**, 121—126 (1965).

338. Nishmi, M., Keller, R.: Titration of vaccinia virus and its neutralizing antibody by the plaque technique. Nature **193**, 150—151 (1962).

339. Nowakowski, M., Bauer, W., Kates, J.: Characterization of a DNA-binding phosphoprotein from vaccinia virus replication complex. Virol. **86**, 217—225 (1978).

340. Nuss, D. L., Paoletti, E.: Methyl group analysis of virion-associated high-molecular weight RNA synthesized *in vitro* by purified vaccinia virus. J. Virol. **23**, 110—116 (1977).

341. Obert, G., Tripier, F., Guir, J.: Arginine requirement for late mRNA transcription of vaccinia virus in KB cells. Biochem. Biophys. Res. Comm. **44**, 362—367 (1971).

342. OBIJESKI, J. F., PALMER, E. L., GAFFORD, L. G., RANDALL, C. C.: Polyacrylamide gel electrophoresis of fowlpox and vaccinia virus proteins. Virol. **51**, 512—516 (1973).

343. ODA, K., JOKLIK, W. K.: Hybridization and sedimentation studies on early and late vaccinia mRNA. J. mol. Biol. **27**, 395—419 (1967).

344. OGIER, G., CHARDONNET, Y., GASSOLO, L.: Role of lysosomes during infection with Shope fibroma virus of primary rabbit kidney tissue culture cells. J. gen. Virol. **22**, 249—253 (1974).

345. OKI, T., FUJIWARA, Y., HEIDELBERGER, C.: Utilization of host-cell DNA by vaccinia virus replicating in HeLa cells irradiated intranuclearly with tritium. J. gen. Virol. **13**, 401—413 (1971).

346. OLGIATI, D., POGO, B. G. T., DALES, S.: Biogenesis of vaccinia: specific inhibition of rapidly labeled host DNA in vaccinia inoculated cells. Virol. **71**, 325—335 (1976).

347. OVERMAN, J. R., TAMM, I.: Multiplication of vaccinia virus in the chorioallantoic membrane *in vitro*. Virol. **3**, 173—184 (1957).

348. PADGETT, B. L., TOMPKINS, J. K. N.: Conditional lethal mutants of rabbitpox virus. III. Temperature-sensitive *(ts)* mutants; physiological properties, complementation and recombination. Virol. **36**, 161—167 (1968).

349 PAOLETTI, E..: *In vitro* synthesis of a high molecular weight virion-associated RNA by vaccinia. J. biol. Chem. **252**, 866—871 (1977a).

350. PAOLETTI, E.: High molecular weight virion-associated RNA of vaccinia. A possible precursor to 8 to 12S mRNA. J. biol. Chem. **252**, 872—877 (1977b).

351. PAOLETTI, E., MOSS, B.: Deoxyribonucleic acid-dependent nucleotide phosphohydrolase activity in purified vaccinia virus. J. Virol. **10**, 866—868 (1972a).

352. PAOLETTI, E., MOSS, B.: Protein kinase and specific phosphate acceptor proteins associated with vaccinia virus cores. J. Virol. **10**, 417—424 (1972b).

353. PAOLETTI, E., ROSEMOND-HORNBEAK, H., MOSS, B.: Two nucleic acid-dependent nucleoside triphosphate phosphohydrolases from vaccinia virus. Purification and characterization. J. biol. Chem. **249**, 3273—3280 (1974).

354. PAOLETTI, E., MOSS, B.: Two nucleic acid-dependent nucleoside triphosphate phosphohydrolases from vaccinia virus. Nucleotide substrate and polynucleotide cofactor specificities. J. biol. Chem. **249**, 3281—3286 (1974).

355. PAOLETTI, E., GRADY, L. J.: Transcriptional complexity of vaccinia virus *in vivo* and *in vitro*. J. Virol. **23**, 608—615 (1977).

356. PAOLETTI, E., LIPINSKAS, B. R.: Soluble endoribonuclease activity from vaccinia virus: specific cleavage of virion-associated high-molecular-weight RNA. J. Virol. **26**, 822—824 (1978a).

357. PAOLETTI, E., LIPINSKAS, B. R.: The role of ATP in the biogenesis of vaccinia virus mRNA *in vitro*. Virol. **87**, 317—325 (1978b).

358. PAOLETTI, E., LIPINSKAS, B. R., PANICALI, D.: Capped and polyadenylated low-molecular-weight RNA synthesized by vaccinia virus *in vitro*. J. Virol. **33**, 208—219 (1980).

359. PARKHURST, J., PETERSON, A. R., HEIDELBERGER, C.: Breakdown of HeLa cell DNA mediated by vaccinia virus. Proc. Nat. Acad. Sci. U.S.A. **70**, 3200—3204 (1973).

360. PARKHURST, J. R., HEIDELBERGER, C.: Rapid lysis of vaccinia virus DNA on neutral sucrose gradients with release of intact DNA. Anal. Biochem. **71**, 53—59 (1976).

361. PARR, R. P., BURNETT, J. W., GARON, C. F.: Structural characterization of the *Molluscum contagiosum* virus genome. Virol. **81**, 247—256 (1977).

362. PAYNE, L. E.: Polypeptide composition of extracellular enveloped vaccinia virus. J. Virol. **27**, 28—37 (1978).

363. PAYNE, L. G.: Identification of the vaccinia hemagglutinin polypeptide from a cell system yielding large amounts of extracellular enveloped virus. J. Virol. **31**, 147—155 (1979).

364. PAYNE, L. G., NORRBY, E.: Presence of hemagglutinin in the envelope of extracellular vaccinia virus particles. J. gen. Virol. **32**, 63—72 (1976).

365. PAYNE, L. G., NORRBY, E.: Adsorption and penetration of enveloped and naked vaccinia virus particles. J. Virol. **27**, 19—27 (1978).

366. PEDRALI-NOY, G., WEISSBACH, A.: Evidence of a repetitive sequence in vaccinia virus DNA. J. Virol. **24**, 406—407 (1977).

367. PELHAM, H. R. B.: Use of coupled transcription and translation to study mRNA production by vaccinia cores. Nature **269**, 532—534 (1977).

368. PENNINGTON, T. H.: Vaccinia virus morphogenesis: a comparison of virus-induced antigens and polypeptides. J. gen. Virol. **19**, 65—79 (1973).

369. PENNINGTON, T. H.: Vaccinia virus polypeptide synthesis: sequential appearance and stability of pre- and post-replicative polypeptides. J. gen. Virol. **25**, 433—444 (1974).

370. PENNINGTON, T. H.: Effect of 5-bromodeoxyuridine on vaccinia virus-induced polypeptide synthesis: selective inhibition of the synthesis of some post-replicative polypeptides. J. Virol. **18**, 1131—1133 (1976).

371. PENNINGTON, T. H., FOLLET, E. A. C., SZILAGYI, J. F.: Events in vaccinia-virus infected cells following the reversal of the antiviral action of rifampicin. J. gen. Virol. **9**, 225—237 (1970).

372. PENNINGTON, T. H., FOLLET, E. A. C.: Vaccinia virus replication in enucleate BSC-1 cells: particle production and synthesis of viral DNA and proteins. J. Virol. **13**, 488—493 (1974).

373. PERRIN, L., ZINKERNAGEL, R., OLDSTONE, M. B. A.: Immune specific killing of vaccinia virus targets by PBLs of humans following vaccination. Fed. Proc. **36**, 1228 (1978). (Abstract.)

374. PETERS, D.: Morphology of resting vaccinia virus. Nature **178**, 1453—1455 (1956).

375. PETERS, D., MÜLLER, G.: The fine structure of the DNA-containing core of vaccinia virus. Virol. **21**, 266—269 (1963).

376. PETERS, D., MÜLLER, G., BUTTNER, D.: The fine structure of paravaccinia viruses. Virol. **23**, 609—611 (1964).

377. PFAU, C. J., McCREA, J. F.: Some unusual properties of vaccinia virus deoxyribonucleic acid. Biochim. biophys. Acta **55**, 271—272 (1962a).

378. PFAU, C. T., McCREA, J. F.: Release of DNA from vaccinia virus by 2-mercaptoethanol and pronase. Nature **194**, 894—895 (1962b).

379. PFAU, C. J., McCREA, J. F.: Studies on the deoxyribonucleic acid of vaccinia virus. III. Characterization of DNA isolated by different methods and its relation to virus structure. Virol. **21**, 425—435 (1963).

380. PITKANEN, A., McAUSLAN, B., HEDGPETH, J., WOODSON, B.: Induction of poxvirus ribonucleic acid polymerases. J. Virol. **2**, 1363—1367 (1968).

381. PLANTEROSE, D. N., NISHIMURA, C., SALZMAN, N. P.: The purification of vaccinia virus from cell cultures. Virol. **18**, 294—301 (1962).

382. POGO, B. G. T.: Elimination of naturally occurring cross-links in vaccinia virus DNA after viral penetration into cells. Proc. Nat. Acad. Sci. U.S.A. **74**, 1739—1742 (1977).

383. POGO, B. G. T.: Formation of cross-linked molecules during vaccinia DNA replication. Intervirol. **11**, 196—200 (1979).

384. POGO, B. G. T.: Terminal cross-linking of vaccinia DNA strands by an *in vitro* system. Virol. **100**, 339—347 (1980a).

385. POGO, B. G. T.: Changes in parental vaccinia virus DNA after viral penetration into cells. Virology **101**, 520—524 (1980b).

386. POGO, B. G. T., DALES, S.: Two deoxyribonuclease activities within purified vaccinia virus. Proc. Nat. Acad. Sci. U.S.A. **63**, 820—827 (1969a).

387. POGO, B. G. T., DALES, S.: Regulation of the synthesis of nucleotide phosphohydrolase and neutral deoxyribonuclease. Proc. Nat. Acad. Sci. U.S.A. **63**, 1297—1303 (1969b).

388. POGO, B. G. T., DALES, S.: Biogenesis of vaccinia separation of early stages from maturation by means of hydroxyurea. Virol. **43**, 144—151 (1971).

389. POGO, B. G. T., DALES, S., BERGOIN, M., ROBERTS, D. W.: Enzymes associated with an insect poxvirus. Virol. **43**, 306—309 (1971).

390. Pogo, B. G. T., Dales, S.: Biogenesis of poxviruses: inactivation of host DNA polymerase by a component of the invading inoculum particle. Proc. Nat. Acad. Sci. U.S.A. **70**, 1726—1729 (1973).
391. Pogo, B. G. T., Dales, S.: Biogenesis of poxviruses: further evidence for inhibition of host and virus DNA synthesis by a component of the invading inoculum particle. Virol. **58**, 377—386 (1974).
392. Pogo, B. G. T., Katz, J. R., Dales, S.: Biogenesis of poxviruses: synthesis and phosphorylation of a basic protein associated with the DNA. Virol. **64**, 531—543 (1975).
393. Pogo, B. G. T., O'Shea, M. T.: Further characterization of deoxyribonucleases from vaccinia virus. Virol. **77**, 55—66 (1977).
394. Pogo, B. G. T., O'Shea, M. T.: The mode of replication of vaccinia virus DNA. Virol. **86**, 1—8 (1978).
394a. Pogo, B. G. T., O'Shea, M. T.: Studies on vaccinia DNA replication. 2nd Cold Spring Harbor Poxvirus Iridiovirus Workshop 1979. (Abstract.)
395. Polisky, B., Kates, J.: Interaction of vaccinia DNA-binding proteins with DNA *in vitro*. Virol. **69**, 143—147 (1976).
396. Postlethwaite, R.: Molluscum contagiosum. Arch. environ. Health **21**, 432—452 (1970).
397. Postlethwaite, R., Watt, J. A.: The virus of Molluscum contagiosum and its adsorption to mouse embryo cells in culture. J. gen. Virol. **1**, 269—280 (1967).
398. Prescott, D. M., Kates, J., Kirkpatrick, J. B.: Replication of vaccinia virus DNA in enucleated L-cells. J. mol. Biol. **59**, 505—508 (1971).
399. Prose, P. H., Friedman-Kein, A. E., Vilček, J.: Molluscum contagiosum virus in adult human skin cultures. Am. J. Path. **55**, 349—366 (1969).
400. Randall, C. C., Gafford, L. G., Soehner, R. L., Hyde, J. M.: Physicochemical properties of fowlpox virus deoxyribonucleic acid and its anomalous infectious behavior. J. Bact. **91**, 95—100 (1966).
401. Reed, L. J., Muench, H.: A simple method of estimating fifty percent-endpoints. Am. J. Hyg. **27**, 493—497 (1938).
402. Roberts, D. W.: A poxlike virus from *Amsacta moorei (Lepidoptera: Arctiidae)*. J. Path. **12**, 141—143 (1968).
403. Robinow, C. F.: A note on stalked forms of viruses. J. gen. Microbiol. **4**, 242 to 244 (1950).
404. Robinson, H., jr., Prose, P. H., Friedman-Kien, A. E., Neistein, S., Vilček, J.: The Molluscum contagiosum virus in chick embryo cultures: an electron microscopic study. J. invest. Dermatol. **52**, 51—56 (1969).
405. Roening, G., Holowczak, J. A.: Evidence for the presence of RNA in the purified virions of vaccinia virus J. Virol. **14**, 707—708 (1974).
406. Rosemond-Hornbeak, H., Paoletti, E., Moss, B.: Single-stranded deoxyribonucleic acid-specific nuclease from vaccinia virus. Purification and characterization. J. biol. Chem. **249**, 3287—3291 (1974a).
407. Rosemond-Hornbeak, H., Moss, B.: Single-stranded deoxyribonucleic acid-specific nuclease from vaccinia virus. Endonucleolytic and exonucleolytic activities. J. biol. Chem. **249**, 3294—3296 (1974b).
408. Rosemond-Hornbeak, H., Moss, B.: Inhibition of host protein synthesis by vaccinia virus: fate of cell mRNA and synthesis of small poly (A)-rich polyribonucleotides in the presence of Actinomycin D. J. Virol. **16**, 34—42 (1975).
409. Rosenkranz, H. S., Rose, H. M., Morgan, C., Hsu, K. C.: Effect of hydroxyurea on virus development. II. Vaccinia virus. Virol. **28**, 510—519 (1966).
410. Rouget, P., Parodi, A., Blangy, D., Cuzin, F.: Origin of polyoma virus-associated endonuclease. J. Virol. **20**, 9—13 (1976).
411. Sahu, S. P., Minocha, H. C.: Fibroma virus-induced acid deoxyribonuclease in rabbit kidney cells. Arch. ges. Virusforsch. **44**, 14—22 (1974).
412. Salzman, N. P., Sebring, E. D.: Sequential formation of vaccinia virus proteins and viral DNA replication. J. Virol. **1**, 16—23 (1967).

413. SAMBROOK, J. F., McCLAIN, M. E., EASTERBROOK, K. B., McAUSLAN, B. R.: A mutant of rabbitpox virus defective at different stages of its multiplication in three cell types. Virol. **26**, 738—745 (1965).
414. SAMBROOK, J. F., PADGETT, B. L., TOMKINS, J. K. N.: Conditional lethal mutants of rabbitpox virus. I. Isolation of host cell-dependent and temperature-dependent mutants. Virol. **28**, 592—599 (1966).
415. SAMBROOK, J., SHATKIN, A. J.: Polynucleotide ligase activity in cells infected with simian virus 40 polyoma virus and vaccinia virus. J. Virol. **4**, 719—726 (1969).
416. SAROV, I., JOKLIK, W. K.: Studies on the nature and location of the capsid polypeptides of vaccinia virions. Virol. **50**, 579—592 (1972a).
417. SAROV, I., JOKLIK, W. K.: Characterization of intermediates in the uncoating of vaccinia virus DNA. Virol. **50**, 593—602 (1972b).
418. SAROV, I., JOKLIK, W. K.: Isolation and characterization of intermediates in vaccinia virus morphogenesis. Virol. **52**, 223—233 (1973).
419. SCHARFF, M. D., SHATKIN, A. J., LEVINTOW, L.: Association of newly formed viral protein with specific polyribosomes. Proc. Nat. Acad. Sci. U.S.A. **50**, 686—694 (1963).
420. SCHROM, M., BABLANIAN, R.: Inhibition of protein synthesis by vaccinia virus. I. Characterization of an inhibited cell-free protein-synthesizing system from infected cells. Virol. **99**, 319—328 (1979).
421. SCHÜMPERLI, D., PETERHANS, E., WYLER, R.: Permeability changes of plasma and lysosomal membranes in HeLa cells infected with rabbit poxvirus. Arch. Virol. **58**, 203—212 (1978).
422. SCHÜMPERLI, D., McFADDEN, G. WYLER, R., DALES, S.: Location of a new endonuclease restriction site associated with a temperature-sensitive mutation of vaccinia virus. Virol. **101**, 281—285 (1980).
423. SCHWARTZ, J., DALES, S.: Biogenesis of poxviruses: identification of four enzyme activities within purified Yaba tumor virus. Virol. **45**, 797—801 (1971).
424. SHAND, J. H., GIBSON, P., GREGORY, D. W., COOPER, R. J., KEIR, H. M., POSTLETHWAITE, R.: *Molluscum contagiosum* — a defective poxvirus? J. gen. Virol. **33**, 281—295 (1976).
425. SHARP, D. G.: Quantitative use of the electron microscope in virus research. Methods and recent results of particle counting. Lab. Invest. **14**, 831—863 (1965).
426. SHARP, D. G., BEARD, J. W.: Counts of virus particles by sedimentation on agar and electron micrography. Proc. Soc. exp. Biol. Med. **81**, 75—79 (1952).
427. SHARP, D. G., SMITH, K. O.: Rapid adsorption of vaccinia virus on tissue culture cells by centrifugal force. Proc. Soc. exp. Biol. Med. **104**, 167—169 (1960).
428. SHARP, D. G., McGUIRE, P. M.: Spectrum of physical properties among the virions of a whole population of vaccinia virus particles. J. Virol. **5**, 275—281 (1970).
429. SHATKIN, A. J.: Actinomycin D and vaccinia virus infection of HeLa cells. Nature **199**, 357—358 (1963).
430. SHATKIN, A. J., SEBRING, E. D., SALZMAN, N. P.: Vaccinia virus directed RNA: its fate in the presence of Actinomycin. Science **148**, 87—90 (1965).
431. SHEEK, M. R., MAGEE, W. E.: An autoradiographic study of the intracellular development of vaccinia virus. Virol. **15**, 146—163 (1961).
432. SHELDON, R., KATES, J., KELLEY, D. E., PERRY, R. P.: Polyadenylic acid sequences on 3' termini of vaccinia mRNA and mammalian nuclear and mRNA. Biochem. **11**, 3829—3834 (1972).
433. SHELDON, R., KATES, J.: Mechanism of poly A synthesis by vaccinia virus. J. Virol. **14**, 214—224 (1974).
434. SHELUKHINA, E. M., MARENNIKOVA, S. S., SHENKMAN, L. S., FROLTSOVA, A. E.: Variola virus strains of 1960—1975: the range of intraspecies variability and relationship between properties and geographic origin. Acta Virol. **23**, 360—366 (1979).
435. SHIDA, H., TANABE, K., MATSUMOTO, S.: Mechanism of virus occlusion into A-type inclusion during poxvirus infection. Virol. **76**, 217—233 (1977).

436. SILVER, M., McFADDEN, G., WILTON, S., DALES, S.: Biogenesis of poxviruses: role for the DNA-dependent RNA polymerase II of the host during expression of late functions. Proc. Nat. Acad. Sci. U.S.A. **76**, 4122—4125 (1979).
437. SINGH, S. B., SMITH, J. W., RAWLS, W. E., TEVETHIA, S. S.: Demonstration of cytotoxic antibodies in rabbits bearing tumors induced by Shope fibroma virus. Infect. Immun. **5**, 352—358 (1972).
438. SMADEL, J. E., LAVIN, G. I.: Some constituents of elementary bodies of vaccinia. J. exp. Med. **71**, 373—389 (1940).
439. SMADEL, J. E., HOAGLAND, C. L.: Elementary bodies of vaccinia. Bact. Revs. **6**, 79—110 (1942).
440. SMITH, K. O., SHARP, D. G.: Interaction of virus with cells in tissue cultures. I. Adsorption on and growth of vaccinia virus in L cells. Virol. **11**, 519—532 (1960).
441. SOEKAWA, M., MORIGUCHI, R., MORITA, C., KITAMURA, T., TANAKA, Y.: Electron-microscopical observations on the development of vaccinia, cowpox and monkey-pox viruses in pig skin. Zbl. Bakt. Hyg., I. Abt. Orig. **A237**, 425—443 (1977).
442. SOLOSKI, M. J., ESTEBAN, M., HOLOWCZAK, J. A.: DNA binding proteins in the cytoplasm of vaccinia virus-infected mouse L cells. J. Virol. **25**, 263—273 (1978).
443. SPENCER, E., LORING, D., HURWITZ, J., MONROY, G.: Enzymatic conversion of 5'-phosphate-terminated RNA to 5'-di- and triphosphate-terminated RNA. Proc. Natl. Acad. Sci. U.S.A. **75**, 4793—4797 (1978).
444. SPENCER, E., LESING, D., HURWITZ, J.: The role of ATP in RNA synthesis by permeabilized vaccinia virus. 2nd Cold Spring Harbor Poxvirus Iridovirus Workshop, 1979.
445. STERN, W., DALES, S.: Biogenesis of vaccinia: concerning the origin of the envelope phospholipids. Virol. **62**, 293—306 (1974).
446. STERN, W., DALES, S.: Biogenesis of vaccinia: relationship of the envelope to virus assembly. Virol. **75**, 242—255 (1976a).
447. STERN, W., DALES, S.: Biogenesis of vaccinia: isolation and characterization of a surface component that elicits antibody suppressing infectivity and cell-cell fusion. Virol. **75**, 323—341 (1976b).
448. STERN, W., POGO, B. G. T., DALES, S.: Biogenesis of poxviruses: analysis of the morphogenetic sequence using a conditional lethal mutant defective in envelope self-assembly. Proc. Nat. Acad. Sci. U.S.A. **74**, 2162—2166 (1977).
449. STEVENIN, J., PETER, R., KIRN, A.: Action des temperatures supraoptimales sur la transcription du virus vaccinal. Biochim. biophys. Acta **199**, 363—372 (1969).
450. STOECKENIUS, W., PETERS, D.: Untersuchungen am Virus der Viriolavaccine, IV. Mitt.: Die Morphologie des Innenkörpers. Zschr. Naturforsch. **106**, 77—80 (1955).
451. STOKES, G. V.: High voltage electron microscope study of the release of vaccinia virus from whole cells. J. Virol. **18**, 636—643 (1976).
452. STOLTZ, D. B., SUMMERS, M. D.: Observations on the morphogenesis and structure of a hemocytic poxvirus in the midge *Chironomus attenuatus*. J. Ultrast. Res. **40**, 581—598 (1972).
453. SUBAK-SHARPE, J. H., PENNINGTON, T. H., SZILAGYI, J. F., TIMBURY, M. C., WILLIAMS, J. F.: The effect of rifampicin on mammalian viruses and cells. In: RNA-polymerase and transcription. 1st International Lepetit Colloquium, Florence, 261—286. Amsterdam-London: North-Holland 1969a.
454. SUBAK-SHARPE, J. H., TIMBURY, M. C., WILLIAMS, J. F.: Rifampicin inhibits the growth of some mammalian viruses. Nature **222**, 341—345 (1969b).
455. TANIGAKI, T., KATO, S.: Autoradiographic studies on *Molluscum contagiosum* using ^{3}H-thymidine. Biken J. **10**, 41—44 (1967).
456. TEMIN, H.: Mechanism of cell transformation by RNA tumor viruses. Ann. Rev. Microbiol. **25**, 609—648 (1971).
457. TOMPKINS, W. A. F., WALKER, D. L., HINZE, H. C.: Cellular deoxyribonucleic acid synthesis and loss of contact inhibition in irradiated and contact-inhibited cell cultures infected with fibroma virus. J. Virol. **4**, 603—609 (1969).

458. TOMPKINS, W. A. F., ZARLING, J. M., RAWLS, W. E.: *In vivo* assessment of cellular immunity to vaccinia virus: contribution of lymphocytes and macrophages. Infect. Immun. **2**, 783—790 (1970a).

459. TOMPKINS, W. A. F., ADAMS, C., RAWLS, W. E.: An *in vitro* measure of cellular immunity to fibroma virus. J. Immunol. **104**, 502—510 (1970b).

460. TOMPKINS, W. A. F., CROUCH, N. A., TEVETHIA, S. S., RAWLS, W. E.: Characterization of surface antigen on cells infected by fibroma virus. J. Immunol. **105**, 1181—1189 (1970c).

461. TOMPKINS, W. A. F., SCHULTZ, R. M.: Cytotoxic antibody response to tumors induced in adult and newborn rabbits by fibroma virus. Infect. Immun. **6**, 591—599 (1972).

462. TOMPKINS, W. A. F., SCHULTZ, R. M., RAMA RAO, G. V.: Depressed cell-mediated immunity in newborn rabbits bearing fibroma virus-induced tumors. Infect. Immun. **7**, 613—619 (1973).

463. TOMPKINS, W. A. F., RAMA RAO, G. V.: Defective macrophage immunity in newborn rabbits with fibroma virus-induced tumors. J. Reticuloendothel. Soc. **23**, 161—166 (1978).

464. TSUCHIYA, Y., TAGAYA, I.: Isolation of multinucleated giant cell-forming and hemagglutinin-negative variant of variola virus. Arch. ges. Virusforsch. **39**, 292—295 (1972).

465. TSUCHIYA, Y., TAGAYA, I.: Rescue of host-dependent conditional lethal mutants of vaccinia and rabbitpox viruses by Yaba virus. Arch. Virol. **55**, 341—345 (1977).

466. TSUCHIYA, Y., TAGAYA, I.: Plaque formation by a host range mutant of vaccinia virus in non-permissive cells coinfected with Yaba virus. J. gen. Virol. **43**, 193—202 (1979).

467. TSURUHARA, T.: Immature particle formation of Yaba poxvirus studied by electron microscopy. J. Nat. Cancer Inst. **47**, 549—554 (1971).

468. TSURUHARA, T., TSURUHARA, A.: Further studies on the development of Yaba poxvirus in cell culture. Arch. ges. Virusforsch. **43**, 119—134 (1973).

469. TURNER, A., BAXBY, D.: Structural polypeptides of *Orthopoxvirus*: their distribution in various members and location within the virion. J. gen. Virol. **45**, 537—545 (1979).

470. TUTAS, D. J., PAOLETTI, E.: Purification and characterization of core-associated polynucleotide 5'-triphosphatase from vaccinia virus. J. biol. Chem. **252**, 3092—3098 (1977).

471. TUTAS, D. J., PAOLETTI, E.: Synthesis of polynucleotide 5'-triphosphatase in vaccinia virus-infected HeLa cells. J. Virol. **25**, 37—41 (1978).

472. UEDA, Y., ITO, M., TAGAYA, I.: A specific surface antigen induced by poxvirus. Virol. **38**, 180—182 (1969).

473. URISHIBARA, T., FURNICHI, Y., NISHIMURA, C., MIURA, K.: A modified structure at the 5'-terminal of mRNA of vaccinia virus. FEBS Lett. **49**, 385—389 (1975).

474. VALDIMARSSON, H., AGNARSDOTTIR, G., LACHMANN, P. J.: Lysis mediated by T cells and restricted by H-2 antigen of target cells infected with vaccinia virus. Nature **255**, 552—556 (1975).

475. VALLEJO-FREIRE, A., BRUNNER, A., BECAK, W.: Vaccinia virus multiplication in rabbit-kidney cell cultures. Aspects of the evolution cycle. Mem. Inst. Butantan **28**, 275—302 (1957—1958).

476. VARICH, N. L., SYCHOVA, I. V., ANTONOVA, T. P., CHERNOS, V. I.: Transcription of both DNA strands of vaccinia virus genome *in vivo*. Virol. **96**, 412—420 (1979).

477. VOLPINO, G.: Corpuscoli mobili, specifici dell-infezione vaccinia nell-epitelio corneale del conigli. Rivista di Igiene e di Sanita Pubblica **18**, 737—747 (1907).

478. VON BORRIES, B., RUSKA, E., RUSKA, H.: Bakterie und Virus in übermikroskopischer Aufnahme (mit einer Einführung in die Technik des Übermikroskops). Klin. Wschr. **17**, 921—925 (1938).

479. VON PASCHEN, E.: Was wissen wir über den Vakzineerreger. Münch. med. Wschr. **49**, 2391—2393 (1906).

480. VREESWIJK, J., LEENE, W., KALSBEEK, G. L.: Early interactions of the virus *Molluscum contagiosum* with its host cell. Virus-induced alterations in the basal and suprabasal layers of the epidermis. J. ultrastruc. Res. **54**, 37—52 (1976).

481. VREESWIJK, J., LEENE, W., KALSBEEK, G. L.: Early host cell-*Molluscum contagiosum* virus interactions. II. Viral interactions with the basal epidermal cells. J. invest. Dermatol. **69**, 249—256 (1977).

482. WEBER, L. A., HICKEY, E. D., NUSS, D. L., BAGLIONI, C.: 5′ terminal 7-methyl-guanosine and mRNA function: Influence of potassium concentration on translation *in vitro*. Proc. Nat. Acad. Sci. U.S.A. **74**, 3254—3258 (1977).

483. WEI, C. M., MOSS, B.: Methylation of newly synthesized viral messenger RNA by an enzyme in vaccinia virus. Proc. Nat. Acad. Sci. U.S.A. **71**, 3014—3018 (1974).

484. WEI, C. M., MOSS, B.: Methylated nucleotides block 5′-terminus of vaccinia virus mRNA. Proc. Nat. Acad. Sci. U.S.A. **72**, 318—322 (1975).

485. WEINTRAUB, S., DALES, S.: Biogenesis of poxviruses: genetically controlled modifications of structural and functional components of the plasma membrane. Virol. **60**, 96—127 (1974).

486. WEINTRAUB, S., STERN, W., DALES, S.: Biogenesis of vaccinia. Effects of inhibitors of glycosylation on virus-mediated activities. Virol. **78**, 315—322 (1977).

487. WESTWOOD, J. C. N., HARRIS, W. J., ZWARTOUW, H. T., TITMUSS, D. H. J., APPLEYARD, G.: Studies on the structure of vaccinia virus. J. gen. Microbiol. **34**, 67—78 (1964).

488. WESTWOOD, J. C. N., ZWARTOUW, H. T., APPLEYARD, G., TITMUSS, D. H. J.: Comparison of the soluble antigens and virus particle antigens of vaccinia virus. J. gen. Microbiol. **38**, 47—53 (1965).

489. WHITE, H. B., JR., POWELL, S. S., GAFFORD, L. G., RANDALL, C. C.: The occurrence of squalene in lipid of fowlpox virus. J. biol. Chem. **243**, 4517—4525 (1968).

490. WILKIN, J. K.: Molluscum contagiosum venereum in a women's outpatient clinic: A venereally transmitted disease. Am. J. Obstet. Gynecol. **128**, 531—535 (1977).

491. WILLIAMS, R. C., FRAZER, D.: Structural and functional differentiation in T2 bacteriophage. Virol. **2**, 289—307 (1956).

492. WILLIAMSON, J. D.: The effect of methylglyoxal bis(guanyl hydrazone) on vaccinia virus replication. Biochem. Biophys. Res. Comm. **73**, 120—126 (1976).

493. WILLIAMSON, J. D., COOKE, B. C.: Argininosuccinate synthetase-lyase in vaccinia virus-infected HeLa and mouse L cells. J. gen. Virol. **21**, 349—357 (1973).

494. WILCOX, W. C., COHEN, G. H.: Soluble antigens of vaccinia-infected mammalian cells. II. Time course of synthesis of soluble antigens and virus structural proteins. J. Virol. **1**, 500—508 (1967).

495. WISHART, F. O., CRAIGIE, J.: Studies on the soluble precipitable substances of vaccinia. III. The precipitin responses of rabbits to the LS antigen of vaccinia. J. exp. Med. **64**, 831—841 (1936).

496. WITTEK, R., MENNA, A., SCHÜMPERLI, D., STOFFEL, S., MÜLLER, H. K., WYLER, R.: Hind III and Sst I restriction sites mapped on rabbit poxvirus and vaccinia virus DNA. J. Virol. **23**, 669—678 (1977).

497. WITTEK, R., MENNA, A., MÜLLER, K., SCHÜMPERLI, D., BOSLEY, P. G., WYLER, R.: Inverted terminal repeats in rabbit poxvirus and vaccinia virus DNA. J. Virol. **28**, 171—181 (1978a).

498. WITTEK, R., MÜLLER, H. K., MENNA, A., WYLER, R.: Length heterogeneity in the DNA of vaccinia virus is eliminated on cloning the virus. FEBS Lett. **90**, 41—46 (1978b).

499. WITTEK, R., KUENZLE, C. C., WYLER, R.: High C + G content in parapoxvirus DNA. J. gen. Virol. **43**, 231—234 (1979).

500. WITTEK, R., HERLYN, M., SCHÜMPERLI, D., BACHMANN, P. A., MAYR, A., WYLER, R.: Genetic and antigenic heterogeneity of different parapoxvirus strains. Intervirol. **13**, 33—41 (1980a).

501. WITTEK, R., BARBOSA, E., COOPER, J. A., GARON, C. F., CHAN, H., MOSS, B.: Inverted terminal repetition in vaccinia virus DNA encodes early mRNAs. Nature **285**, 21—25 (1980b).
502. WOLSTENHOLME, J., WOODWARD, C. G., BURGOYNE, R. D., STEPHEN, J.: Vaccinia virus cytotoxin. Arch. Virol. **53**, 25—37 (1977).
503. WOODSON, B.: Vaccinia mRNA synthesis under conditions which prevent uncoating. Biochem. biophys. Res. Comm. **27**, 169—175 (1967).
504. YOHN, D. S., MARMOL, F. R., OLSEN, R. G.: Growth kinetics of Yaba tumor poxvirus after *in vitro* adaptation to Cercopithecus kidney cells. J. Virol. **5**, 205—211 (1970).
505. ZUCKERMAN, A., RONDLE, C.: The enigma of poxviruses. Nature **276**, 212—213 (1978).
506. ZWARTOUW, H. T.: The chemical composition of vaccinia virus. J. gen. Microbiol. **34**, 115—123 (1964).
507. ZWARTOUW, H. T., WESTWOOD, J. C. N., APPLEYARD, G.: Purification of poxviruses by density gradient centrifugation. J. gen. Microbiol. **29**, 523—529 (1962).
508. ZWARTOUW, H. T., WESTWOOD, J. C. N., HARRIS, W. J.: Antigens from vaccinia virus particles. J. gen. Microbiol. **38**, 39—45 (1965).

Appended References

509. BAROUDY, B. M., MOSS, B.: Purification and characterization of a DNA-dependent RNA polymerase from vaccinia virions. J. biol. Chem. **255**, 4372—4380 (1980).
510. SPENCER, E., SHUMAN, S., HURWITZ, J.: Purification and properties of vaccinia virus DNA-dependent RNA polymerase. J. biol. Chem. **255**, 5388—5395 (1980).
511. WITTEK, R., MOSS, B.: Tandem repeats within the inverted terminal repetition of vaccinia virus DNA. Cell **21**, 277—284 (1980).
512. MORGAN, J., ROBERTS, B.: Viral gene expression in vaccinia infected L-cells. Abstract 3rd Cold Spring Harbor Workshop on Poxviruses and Iridoviruses, 1980.
513. MAHR, A., ROBERTS, B.: Organization and expression of some cloned EcoRI vaccinia virus fragments. Abstract 3rd Cold Spring Harbor Workshop on Poxviruses and Iridoviruses, 1980.
514. BELLE ISLE, H., VENKATESAN, S., MOSS, B.: Mapping of vaccinia virus early and late polypeptides by cell free translation of mRNA selected by hybridization to cloned DNA fragments. Abstract 3rd Cold Spring Harbor Workshop on Poxviruses and Iridoviruses, 1980.

VIROLOGY
MONOGRAPHS

VIROLOGY MONOGRAPHS continued